D0996859

The New Field Guide to Fungi

The New Field Guide to Fungi

Eric Soothill
&
Alan Fairhurst

FOREWORD BY
H.R.H. The Prince Philip, Duke of Edinburgh

MICHAEL JOSEPH · LONDON

First published in Great Britain by
Michael Joseph Ltd
52 Bedford Square, London, W.C.1.
1978

ISBN 0 7181 1620 8

Filmset by BAS Printers Limited,
Over Wallop, Hampshire, printed by
Sackville Press, Billericay, Essex and
bound by J. M. Dent, Letchworth, Herts

To Gwyneth and Doreen

Contents

The line illustrations appearing between pages 16 and 19 were drawn by Yvonne Skargon.

The British have a reputation for their interest in natural history. The great parks, reserves and botanical gardens in the old colonies are testimony to this characteristic. Yet at home their interest in nature seems to be limited to flowers and birds. In a book on Fungi published in 1909 E.W. Swanton notes that, 'There are books innumerable on common wildflowers . . . but there are very few indeed on common fungi'. Nearly 70 years later, and in an age of even greater emphasis on conservation and nature study the position is much the same.

Now at last Eric Soothill and Alan Fairhurst have produced this 'New Field Guide to Fungi' and I can only hope that it will stimulate a much wider interest in this neglected but fascinating branch of natural history.

Philip

Acknowledgements

The authors are grateful to the following people and organisations for the help they gave in a variety of ways.

In Great Britain:

Mr A. Aldridge
Amenities & Recreation Dept.,
  Metropolitan Borough of Calderdale
Mr Ron Barker
Mr John Blunt
Bolton Museum
Bournemouth Natural Science Society
Mr Clifford Bradley
Dr Wilfred Bradley
Mr Allan Brindle, MSc
Mr Brian Cave
The Lord Clitheroe
Mr Peter Crow
Darlington and Teesdale Naturalists' Club
Mr Fred Earnshaw
Dr Beti Evans
Mr M. J. Fitzherbert-Brockholes, JP
Mr D. G. Gooding
Mr S. E. Greenwood
Dr A. & Miss P. Hall
Mr J. Hardbattle
Mr J. M. Heap
Mr Peter Hill
Mr & Mrs L. Livermore
Mr Frank Murgatroyd
Mr & Mrs Desmond Parish
Mr Roy W. Rhodes
Mr Neil Robinson, MA
Messrs. Synchemicals Ltd., London
Dr George Taylor
Prof D. H. Valentine
Mr Ralph Walker
Wigan & District Field Club

In North America:

Mr & Mrs David G. Dorsey
Mr & Mrs Harry Knighton
Mrs Ruby McAlister
Dr Kent H. McKnight
Dr Orson K. Miller Jnr
Dr D. H. Mitchel
The New York Botanical Gardens
Mr Paul Plante
Mrs Robert 'Kit' Scates
Mr Hank Shank
Dr D. E. Stuntz
Mrs Ellen Trueblood
Dr Ernest Wells

We are especially indebted to Dr Roy Watling, Principal Scientific Officer, Royal Botanic Garden, Edinburgh, not only for preparing a Key to the major genera but for his help in the identification of certain species and the constructive criticism he so willingly gave on reading the original manuscript.

May we also express our thanks to Prof D. H. Valentine for reading the proof.

Introduction

We write this book as non-professional mycologists and direct it towards the amateur in the knowledge that it will help, not only towards a better understanding of the subject, but add to the pleasure already derived from his or her pursuits of natural history. Should the professional also find it of interest, then our efforts will have been the more worthwhile and our reward twofold.

The illustrations represent five years of photography in the field, during which time we have travelled the length and breadth of Britain on several occasions; our journeyings have also taken us to the U.S.A., and into Canada. Being naturalists of many years' standing, the data and general information have been accumulated over a considerable period of time.

What may attract the beginner's attention is the fact that fungi vary so very much in shape and size, often elegant and strikingly beautiful, but sometimes weird and sinister in appearance, and on occasions displaying a variety of colours both vivid and startling. To portray them in their natural surroundings through the media of colour photography and to create an interest for all to share is our aim and pleasure in compiling this book.

Unless otherwise stated within the text, all the species described occur both in Britain and North America.

The taxonomy and nomenclature used are in accordance with the *New Check List—British Agarics and Boleti* (1960) by R. W. G. Dennis, P. D. Orton and F. B. Hora; and an *Index of the Common Fungi of North America* (1975) by Orson K. Miller Jnr and David F. Farr.

Fungi belong to the Vegetable Kingdom, in other words they are plants; but they differ from other plants in being unable to effect photosynthesis as they do not produce chlorophyll, the green colour common to most land plants. Deprived of this, they must lead a parasitic or saprophytic existence and take sustenance from decaying organic matter or from the living cells of other plants or animals.

The fine microscopic filaments (hyphae) of which they are composed are interlaced or tangled into a mass. The majority of fungus species are microscopic and these are not dealt with in this volume.

Fungi are among the major causes of plant diseases and are responsible for a variety of diseases in animals but, along with bacteria, they form an indispensable link in the woodland economy by breaking down organic matter.

It is only in recent years that the importance of fungi has come to be recognised in the field of human medicine. Not all fungi are harmful; some are cultured commercially for the production of drugs, organic acids, enzymes and feed supplements. They are used in the production of certain cheeses and a variety of other foods; the fungi known as yeast are of great economic importance, being used in all parts of the world for baking bread and producing alcoholic beverages by fermentation.

The fruit-bodies which we see above the ground-surface or on the trunks of trees etc., differ greatly in form and size. They are that part of the fungus which produce the spores (the equivalent of seeds in green plants) and on dispersal serve to reproduce the species in other suitable habitats. Spores are microscopic and vary in size according to species but differ from seeds in that they contain neither embryo nor a cellular food reserve and the majority are undivided. In shape they may be described variously as elliptical, sausage-shaped, oval, spherical, spindle-shaped or polygonal; they may be smooth or ornamented. In colour, according to genus or species so may they differ; in mass they can be white, cream, rosy, various shades of brown or even black. Such characters are necessary in the identification of fungi and with the aid of a microscope are not too difficult to ascertain.

To take the spore-print of an agaric, cut the cap from the stipe and place gills down on a piece of white paper; cover with a bell-jar or tumbler and leave for a few hours, say overnight. After this period of time, there should be enough spores deposited to show a definite spore colour print; these spores can be used later for microscopic examination or kept in sealed packets for future reference. A medium-sized agaric will discharge billions of spores, most of which perish under natural conditions.

When a spore settles on a suitable substrate and atmospheric conditions are in harmony, germination takes place. The spore produces a very fine filament or germ-tube which grows outwards, seeking nutrient, soon branching in all directions and anastomosing here and there, eventually forming a complex cobweb-like system. This is the vegetative part of the fungus, collectively called the mycelium and from which the fruit-bodies develop.

In our text, spores may be referred to as being amyloid or dextrinoid. This is a colour test useful in identification and is carried out by using Melzer's reagent which is made up as follows:

Potassium iodide 1·5 grms., iodine crystals 0·5 grms., water 22 grms.

Mix and dissolve these together and finally add chloral hydrate 22 grms.

This mixture is poisonous, should be labelled as such and

kept out of the reach of children. Add a drop of the solution to a spore deposit on a glass slide. If, after a minute or two, the spores become blue to violaceous, the reaction is said to be amyloid; on the other hand, if the spores become reddish-brown, the reaction is said to be dextrinoid.

Rationalisation to the vast array of fungi

The species of fungi described in the following pages, with a few exceptions, belong to the two subdivisions termed, the Basidiomycotina and Ascomycotina, by the expert; these subdivisions are commonly referred to as the Basidiomycetes and Ascomycetes. They are subdivided further according to the method of production and arrangement of the spores. The latter includes the cup-fungi, morels, truffles, earth-tongues, ergot fungus etc., and are characterised by producing spores (usually eight in number, more rarely four or less) within microscopic sacs called asci which are grouped together to form a palisade termed a hymenial layer. When mature the spores are discharged and dispersed elsewhere.

The Basidiomycetes include: agarics, bracket fungi, fairy clubs, stinkhorns, puff-balls, earth-stars and many others. Their spores (two to eight in number but typically four) are born on the outside of club-shaped cells called basidia; these are usually closely packed forming a definite hymenial layer. When ripe, the spores are discharged from the basidium and dispersed in various ways according to the group of Basidiomycetes into which they are placed.

For instance, in the *Gasteromycetales*, the spores form inside the fruit-body and are only discharged when the latter is mature, through holes or cracks in the outer skin as instanced in the puff-balls, earth-balls and earth-stars. In the stinkhorns, the mature spores are to be found in the blackish-olive slime at the top of the fully grown fruit-body. This slime has a foetid odour which attracts muscid flies; they ingest this substance and it sticks to their feet, spores pass through the fly's digestive system and in this manner some of the spores are dispersed. There are also subterranean species whose taste and odour is attractive to invertebrates and the spores are similarly dispersed. The spores of Bird's nest fungi are enclosed in small closed bodies referred to as peridioles which, according to species may or may not be attached by a cord to the inner wall of the peridium or 'nest'. The peridioles are scattered by the action of rain or animals. When the former decay, the spores are released.

In the *Tremellales*, the larger species are mainly gelatinous in consistency becoming shrivelled, rigid and horny when dry, yet regaining their original form when moistened. This group includes such species as Jew's ear and Witch's butter; they are characterised by having septate or forked basidia.

The *Aphyllophorales* is a large and complex group including species which differ conspicuously in shape and habit, but do not have true gills. Many are tough and woody in texture. The group contains species with annual or perennial fruit-bodies. The hymenium may line the inside of tubes; be spread over veins; be borne on spines or warts; or on erect club-like, branched, or coral-like bodies. Such widely differing species as 'Dry-rot fungus', 'Purple Stereum', 'Fairy Club', 'Chantarelle' and 'Birch-bracket' are all members of this group.

The *Agaricales* include the gilled-fungi and boletes. In the former, the hymenium is formed on gills; these are blade-like structures radiating outwardly from the stipe and with vertically arranged sides. Their shape, texture, colour, proximity to one another, and mode of attachment to the stipe are very important in the determination of both genera and species. When young, the gills may be covered by a partial veil which ruptures on expansion of the fruit-body leaving a ring or ring-like zone on the stipe and in some cases appendiculate remains at the cap margin.

In certain agarics and at the early stage, the fruit-body is enclosed in a universal veil, the whole being egg-like in appearance. With growth, this veil ruptures, the remnants often remaining in patches or warts on the cap cuticle and also bag-like or collar-like (the volva) around the stipe base. Owing to the evanescent and fragile nature of these characteristics, their presence or absence should be noted with the fruit-body in situ and before collection, as these features aid greatly in identification.

The texture, position and form of the stipe and cap vary greatly according to species and also play an important part in identification; they may be smooth, silky, hairy, fibrillose, cartilaginous, squamulose, hygrophanous, viscid or dry etc.

The boletes are fleshy, readily decaying fungi with a central stipe, similar to the gill fungi, but having a layer of vertical tubes, opening by pores, instead of radiating gills. Unlike the polypores, these tubes are cleanly separable from the flesh of the cap.

Collecting fungi

When embarking on a fungus foray, it is advisable to carry what might be loosely termed 'tools of the trade'. A large flat-bottomed basket, in which to carry specimens, is essential, plus a few small containers for tiny and fragile species. Plastic bags should not be used, as specimens are liable to sweat and quickly 'go off' if kept in these for even a short period of time.

It must be borne in mind that the prime object is to get one's specimens home or to the laboratory in first-class condition. For obvious reasons different species should not be mixed together; separation can be achieved by wrapping each fruit-body in waxed paper.

A pen-knife or small trowel is useful for digging-up terrestrial species intact and a small pruning saw for extricating specimens from wood. On the spot notes should be taken of as many characteristics in a specimen as possible, especially as to the habitat where it was found, under or on what species of tree it was growing or attached, distinctive odour, change of colour on cutting or breaking, colour and taste of latex if present, whether solitary or in groups etc. In order to examine minute details, a hand lens is also a useful piece of equipment.

Specimens that are not identified in the field should be examined in further detail as soon as possible after collection. If delay is anticipated they should be dried out quickly and completely; in this state, they will retain their microscopic characteristics.

The most rewarding time of year to collect is undoubtedly the late summer and autumn before severe frosts commence; nevertheless, an ardent collector should be rewarded at all other seasons of the year, albeit the number of species will be far less.

In the descriptive text, we give the months of appearance of each fruit-body but it should be remembered that these dates are flexible and only intended to convey to the reader the most likely time of year to find a given species.

Edibility of fungi

Among the many thousands of fungus species with relatively large fruit-bodies, only a few are dangerously or deadly poisonous. Most can be eaten cooked, without ill-effect. These may be classified as follows: gastronomically delightful; very good; fairly good; worthless; indigestible or unpalatable. Before proceeding further it must be pointed out that even the edible species can cause stomach upsets if old, invaded by insect larvae or in a state of decomposition. It is important, therefore, to take only fresh specimens for the table.

Our above judgement of taste is in its broadest sense; for what might be considered very good by a native of Siberia could be positively disgusting to a New York businessman or vice versa. In other words: 'One man's meat may well be another man's poison'.

It is not the purpose of this book to describe in detail the symptoms and effects etc. of fungus poisoning; enough to state that collectors should only eat those species which have been positively identified. Even then it is advisable to eat only small portions of any edible species hitherto untried.

Beware the following 'Old Wives' tales':

It is often said that creatures of the wild will not eat poisonous fungi and that this is a safe way to distinguish between the edible and the harmful or poisonous. Such a statement is most certainly a fallacy, for we often find the deadly poisonous *Amanita phalloides* partly eaten by slugs and snails.

To consider that all spring fungi are edible could also prove fatal, when one remembers that some of the deadly Amanitas occur in that season.

We have read that all mushrooms exuding milk when cut or broken are poisonous, but *Lactarius deliciosus* and *L. sanguifluus* prove this to be false. Both are edible.

Nor are all viscid mushrooms poisonous, for *Boletus elegans* and *Boletus luteus* are good to eat.

Do not be deceived by the statement that all mushrooms lose their poison when boiled in water; some do, but others retain their poisonous properties even after repeated boilings.

Perhaps the best 'old wives' tale' is the one claiming that mushrooms which, while cooking, do not cause a silver spoon to change colour, are edible. We must point out that the Amanitas of which many are very poisonous, do not cause a colour change any more so than *Boletus edulis* which is renowned for its edibility.

Our warning is—*do not take needless risks*. Always seek the advice of a reliable mycologist until such time as personal experience has been gained.

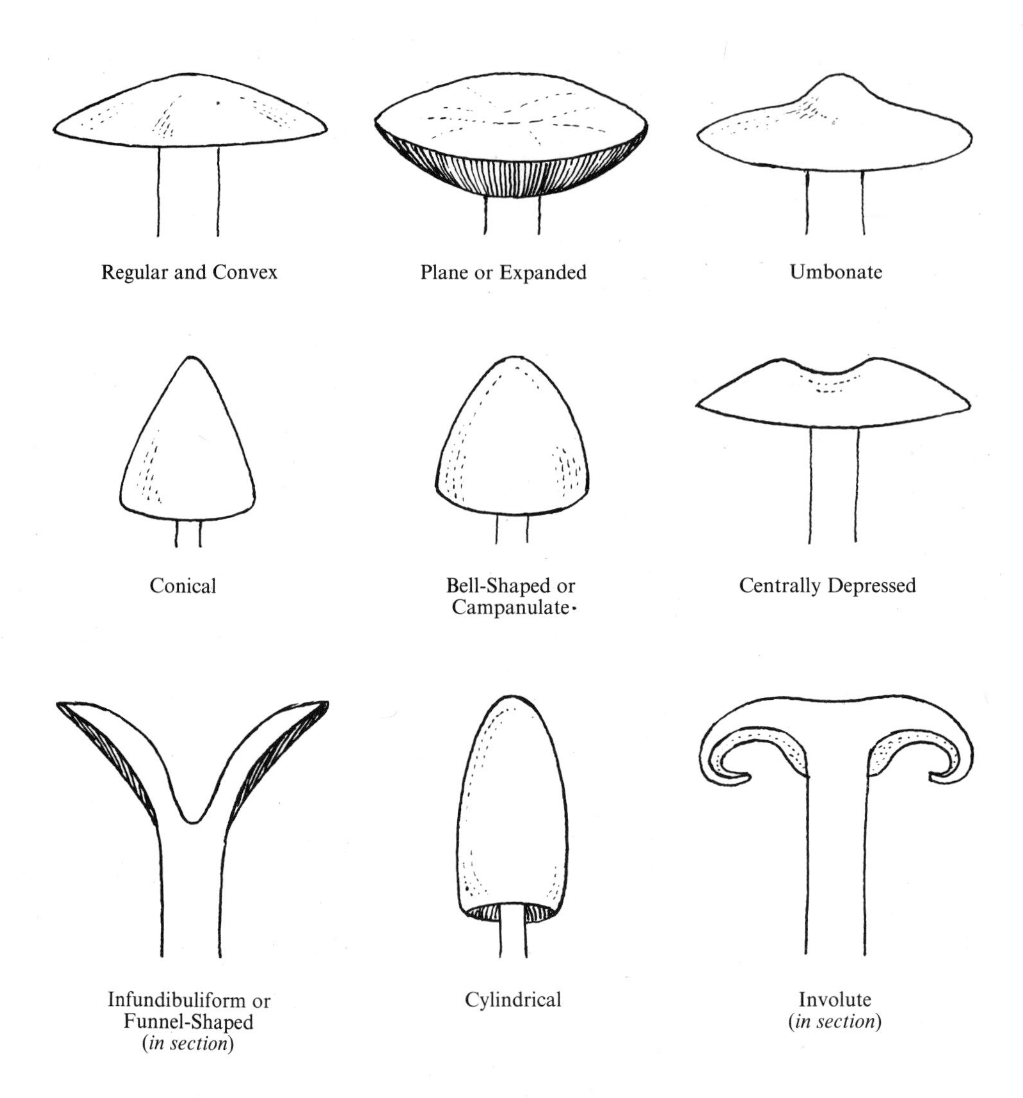

Cap Shapes in Agarics

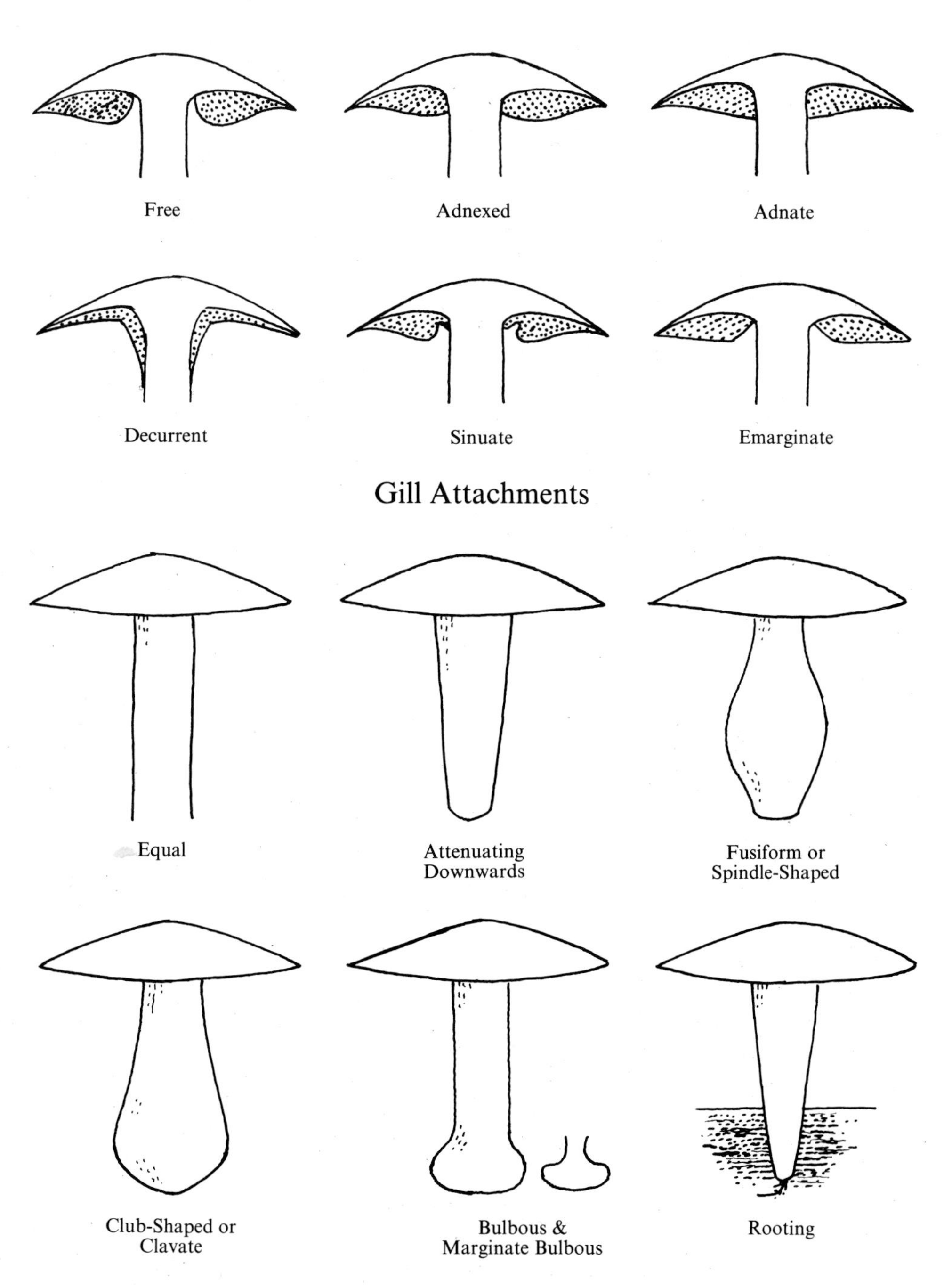
Free
Adnexed
Adnate
Decurrent
Sinuate
Emarginate
Gill Attachments
Equal
Attenuating
Downwards
Fusiform or
Spindle-Shaped
Club-Shaped or
Clavate
Bulbous &
Marginate Bulbous
Rooting
Stipe Characters

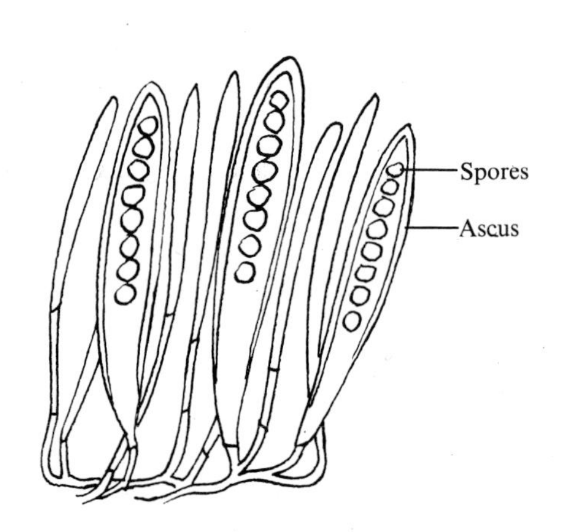

Asci and Sterile Filaments
Highly Magnified

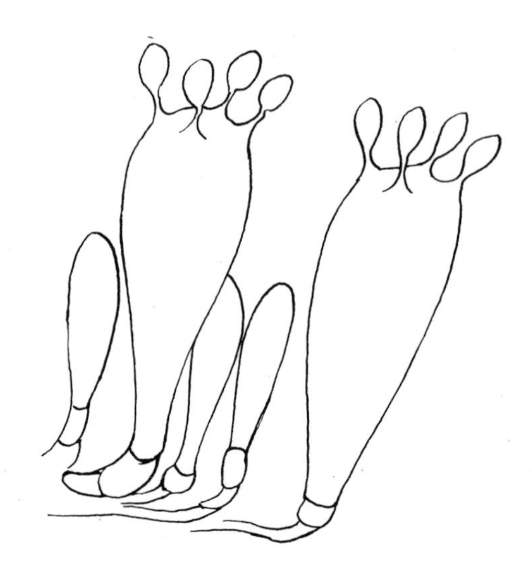

4-Spored Basidia of a Basidiomycete
and
Sterile Cells
(Agaricale Aphyllophorales)

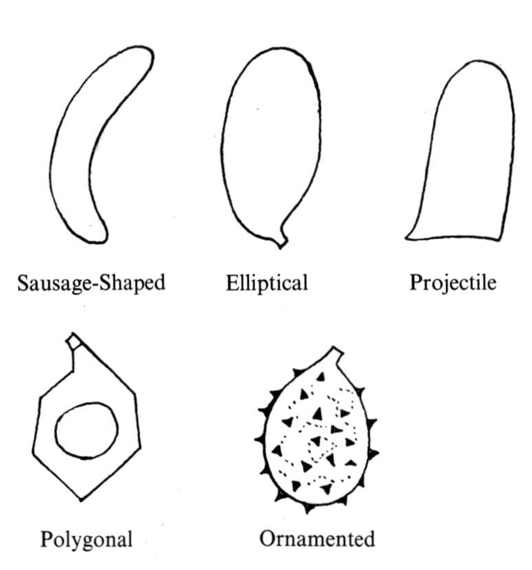

Various Spore Shapes

4-Spored Basidia of a
Gastromycete

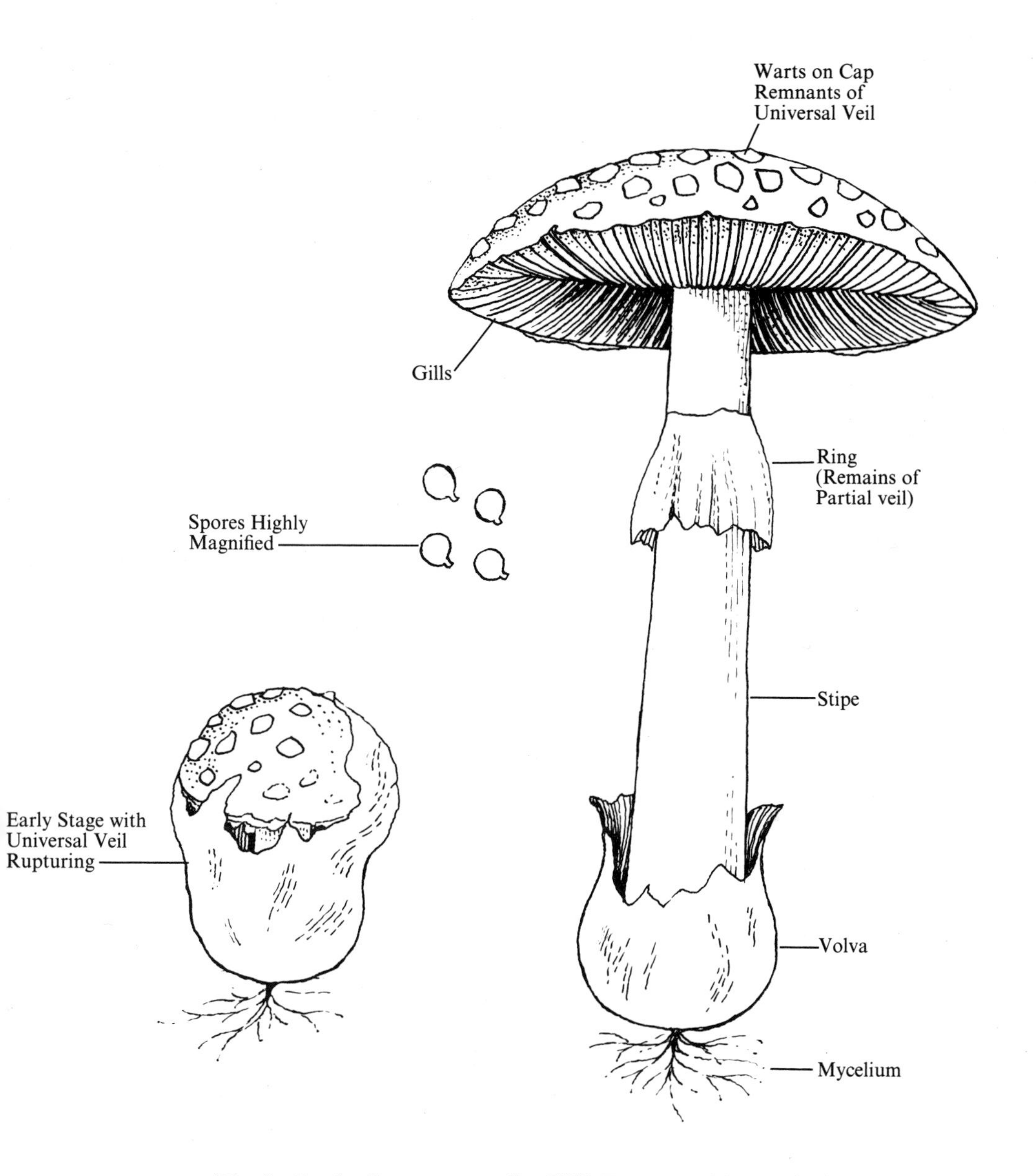

Fruit-Body Structure of a Gill-Fungus (Amanita)

Systematic List of Genera—Agarics & Boleti

Family 1. *CANTHARELLACEAE*

1 Cantharellus
2 Craterellus
3 Leptoglossum
4 Gomphus
= Neurophyllum
5 Plicatura
also Polyozellus

Family 2. *BOLETACEAE*

1 Boletus
includes Tubiporus
Phlebopus
Xerocomus
Pulveroboletus
Suillus
= Ixocomus
Leccinum
2 Tylopilus
3 Porphyrellus
4 Gyroporus
5 Gyrodon
6 Boletinus
7 Strobilomyces
8 Phylloporus
9 Paxillus

Family 3. *GOMPHIDIACEAE*

1 Gomphidius

Family 4. *HYGROPHORACEAE*

1 Hygrophorus
Subgenus i Hygrophorus
= Limacium
Subgenus ii Camarophyllus
Subgenus iii Hygrocybe

Family 5. *PLEUROTACEAE*

1 Pleurotus
includes Pleurocybella
2 Hohenbuehelia
3 Resupinatus
4 Pleurotellus
5 Phyllotopsis
6 Crepidotus
7 Geopetalum
8 Lentinus
9 Lentinellus
10 Panus
11 Panellus
12 Schizophyllum

Family 6. *TRICHOLOMATACEAE*

Tribe (a) Tricholomeae

1 Tricholoma
2 Tricholomopsis
3 Lyophyllum
4 Melanoleuca
= Melaleuca
5 Squamanita

Tribe (b) Clitocybeae

6 Clitocybe
= Omphalia
7 Armillaria
8 Leucopaxillus
9 Cantharellula
10 Hygrophoropsis
11 Laccaria

Tribe (c) Collybieae

12 Collybia
Subgenus i Collybia
Subgenus ii Tephrophana
13 Asterophora
= Nyctalis
14 Oudemansiella
= Mucidula
includes Xerula
15 Flammulina
16 Macrocystidia
17 Clitocybula
18 Dermoloma
19 Pseudohiatula
20 Baeospora
21 Mycena
22 Fayodia
23 Myxomphalia
24 Omphalina
= Omphalia
25 Marasmius
includes Androsaceus
Marasmiellus
26 Micromphale
27 Crinipellis
28 Xeromphalina

Family 7. *CLITOPILACEAE*

1 Clitopilus
2 Lepista
= Rhodopaxillus
3 Rhodocybe
includes Clitopilopsis
4 Rhodotus

Family 8. *RHODOPHYLLACEAE*

1 Entoloma
2 Nolanea
3 Leptonia
4 Eccilia
5 Claudopus

Family 9. *CORTINARIACEAE*

1 Cortinarius
Subgenus i Myxacium
Subgenus ii Phlegmacium
Subgenus iii Sericeocybe
Subgenus iv Cortinarius
= Inoloma
Subgenus v Dermocybe
Subgenus vi Telamonia
includes Hydrocybe
2 Phaeocollybia
3 Leucocortinarius
includes Cortinellus
4 Rozites
5 Phaeolepiota
6 Flocculina
7 Phaeomarasmius
8 Tubaria
9 Gymnopilus
= Fulvidula
includes Flammula
10 Galerina
includes Galera
11 Pholiota
= Dryophila
12 Hebeloma
includes Myxocybe
Hylophila
13 Naucoria
includes Alnicola
Simocybe
Hylophila
14 Inocybe
Subgenus i Inocybe
= Eu-Inocybe
Subgenus ii Clypeus
= genus Astrosporina and genus Clypeus

Family 10. *BOLBITIACEAE*

1 Bolbitius
2 Pluteolus
3 Conocybe
includes Pholiotina
Galerella
Galera
4 Agrocybe
includes Togaria

Family 11. *STROPHARIACEAE*

1 Stropharia
2 Hypholoma
= Naematolma
3 Psilocybe
4 Deconica

Family 12. *COPRINACEAE*

1 Coprinus
includes Pseudocoprinus
Coprinarius
2 Psathyrella
= Drosophila
3 Lacrymaria
4 Panaeolus
includes Coprinarius
Anellaria
5 Panaeolina

Family 13. *AGARICACEAE*

Tribe (a) Agariceae

1 Agaricus
= Pratella
= Psalliota
includes Chitonia
2 Melanophyllum
includes Chlorospora
Glaucospora

Tribe (b) Lepioteae

3 Lepiota
includes Leucocoprinus
Macrolepiota
Leucoagaricus
4 Leucocoprinus
= Hiatula
5 Cystoderma
6 Drosella
= Lepiotella

Family 14. *VOLVARIACEAE*

1 Volvariella
= Volvaria
2 Pluteus

Family 15. *AMANITACEAE*

1 Amanita
includes Amanitopsis
Lepidella
Aspidella
2 Limacella

Family 16. *RUSSULACEAE*

1 Russula
2 Lactarius

Key to the main groups of Larger Fungi

Roy Watling, Royal Botanic Garden, Edinburgh

Only the most important and/or obvious genera are included; for further more critical Keys see a more advanced text, e.g. *British Fungus Flora*, editors—Henderson, Orton & Watling (H.M.S.O.).

Spores borne externally on stalks on a clavate to cylindrical cell: Basidiomycotina—A 1–117

Spores produced within a clavate cylindrical or subglobose cell: Ascomycotina—B 118–130

(Mostly microscopic fungi: only a few are large enough to be confused with the agarics, polypores and their allies, e.g. Morels (Morchella); Dead Man's Finger (Xylosphaera).

NOTE CONCERNING THE USE OF KEY

To those readers who are familiar with a dichotomous type Key there should be no problems as to its use.

For the benefit of those who have never used this type of Key, let us run through one example as a means of explanation.

Suppose we have just collected a specimen of the very well-known Fly Agaric and wish to determine the genus:

Commencing at 1. We see that it agrees with the first description and are thus instructed to proceed to 2.

The basidia is found to be simple and the fruit-body is *not* gelatinous, so agreeing with the first description and on we go to 3.

The fruit-body is not tough and leathery but easily decays, and the spores are produced on the surface of gills. On to 10.

The spores are *not* produced in tubes but on gills, thus indicating a move to 11.

On examining the spore mass we find that it is *not* coloured or only *faintly* so. We continue to 35.

Fruit-body fleshy and easily decaying agrees with our specimen, so on to 36.

It was not growing on other agarics, so on to 37.

The spore-bearing layer is on distinct well-formed gills which indicates proceeding to 39.

The cap is easily separable from the stipe, so we go to 40.

The first part of 40 describes our specimen so we now know it belongs to the genus *Amanita*.

A—Key to major groups based on character of basidium and fruit-body shape

1 Basidia produced in a layer of cells (hymenium) and exposed to the air before the maturity of the spores. (Hymenomyctes) ... 2
  Basidia either produced in a hymenium or in a mass, and until maturity contained within a closed fruit-body. (Gasteromyctes) ... 6

2 Basidia simple; a single cell. (Homobasidiae) ... 3
  Basidia usually septate, or if simple then fruit-body gelatinous and often collapsing to form a skin when dried. (Heterobasidiae) ... 4

3 Fruit-body usually fleshy, soft and easily decaying (putrescent); spores produced on the surface of gills or ridges, or within tubes. (Agaricales) ... (10)
  Fruit-body with spores produced on smooth surfaces, teeth, ridges or plates or if within tubes, then fruit-body tough and leathery. (Aphyllophorales) ... (68)

4 Basidia divided. ... 5
  Basidia simple and apex drawn out into two long necks. (Dacrymycetales)...(111)

5 Basidia divided into two or four cells by vertical cross-walls. (Tremellales)...(115)
  Basidia divided transversely by one to three horizontal cross-walls. (Auriculariales) ...(113)

6 Fruit-body growing beneath soil-surface (hypogeous).
  Hymenogaster & *Rhizopogon*. (False Truffles)
  Fruit-body not growing beneath the soil-surface. ... 7

7 Spores in a slimy mass on a specialised fruit-body arising from an egg-like structure *Phallus* & *Mutinus*. (Stinkhorns)
  Spores powdery at maturity or in small capsules. ... 8

8 Spores powdery at maturity and contained within the fruit-body. ... 9
  Spores enclosed in a small capsule, or group of capsules in a cup-like structure resembling the eggs within the nest of a bird.
  Crucibulum, Cyathus & *Nidula*. (Bird's nest fungi)

9 Spores intermixed with threads within the fruit-body from which they are dispersed through a specialised pore at its apex. (Outer surface intact or flaking away) ... *Lycoperdon*.
  (Outer surface folding back to form a star-like pattern).
  ... *Geastrum*. (Puff-balls and Earth-stars)
  Spores not mixed with threads within the fruit-body and not dispersed through a special structure, but through cracks as the fruit-body weathers.
  ... *Scleroderma*. (Earth-balls)
  (If resembling an unexpanded mushroom, compare with *Endoptychum agaricoides*.)

10 Spores produced on gills, ridges or veins but never in distinct tubes, although gills may become poroid at stem-apex. ... 11
  Spores produced in tubes. ... (62)

11 Spores distinctly coloured in mass and coloured individually under the microscope. ... 12
  Spores not coloured, or only faintly in mass and hyaline under the microscope. ... (35)

12 Spores pinkish. ... 13
  Spores blackish or some shade of brown. ... (17)

13 Stipe laterally attached to the cap or absent. . . . *Claudopus.*
(and some species of *Clitopilus*)
Stipe centrally attached to the cap. . . . 14
14 Stipe with a cup-like structure enveloping the base. . . . *Volvariella.*
Stipe lacking any special structure at its base. . . . 15
15 Gills not attached to the stipe (free), or with part attached to and descending down the stipe (decurrent). . . . 16
Gills attached to the stipe but not descending down the stipe. . . . 17
16 Gills remote to free from the stipe. . . . *Pluteus.*
Gills distinctly attached and descending down the stipe. . . . *Clitopilus.*
(see also *Eccilia*)
17 Gills broadly attached to the stipe (adnate). . . . *Entoloma.*
Gills narrowly attached to the stipe (adnexed). . . . *Leptonia* & *Nolanea.*
18 Stipe laterally attached to the cap. . . . *Crepidotus.*
Stipe centrally attached to the cap. . . . 19
19 Spore-print some shade of brown. . . . 20
Spore-print blackish to purplish black. . . . 28
20 Spore-print bright rust-brown. . . . 21
Spore-print dull clay-brown or ochraceous. . . . 26
21 Stipe with the veil girdling the stem to form a ring or cobweb-like (cortina). . . . 22
Stipe without a ring, or if present then easily lost. . . . 23
22 Stipe with distinct ring or ring-zones. . . . *Pholiota* & related genera.
Stipe with cobweb-like veil or faint filamentous ring-zone.
. . . *Cortinarius* & *Gymnopilus.*
23 Gills attached to the stipe but not descending down the stipe (adnexed to adnate). . . . 24
Gills free of stipe, or distinctly attached to and running down the stipe (decurrent), and then often joined together at the apex of the stipe, or at their base. . . . 25
24 Cap-surface composed of rounded cells. . . . *Conocybe.*
Cap-surface composed of filamentous cells. . . . *Galerina.*
25 Gills free of the stipe and the whole fruit-body very fragile. . . . *Bolbitius.*
Gills attached to and running down the stipe (decurrent), easily separable from the cap-tissue and frequently veined at apex of stipe. . . . *Paxillus.*
26 Cap scaly, fibrillose and roughened. . . . *Inocybe.*
Cap smooth, greasy or viscid. . . . 27
27 Cap-surface composed of rounded cells. . . . *Agrocybe.*
Cap-surface composed of filamentous cells. . . . *Naucoria* & *Hebeloma.*
28 Gills or complete fruit-body becoming liquified. . . . *Coprinus.*
Neither the gills nor the fruit-body collapsing into a slurry of tissue. . . . 29
29 Gills free to remote from the stipe or attached and descending down the stipe (decurrent). . . . 30
Gills attached in some way to the stipe but not descending down the stipe (adnate to adnexed). . . . 31
30 Gills decurrent; stipe possessing a cobweb-like veil.
. . . *Gomphidius* & *Chroogomphus.*

Gills remote or free; stipe usually possessing a persistent ring. (If unexpanding, compare with *Endoptychum agaricoides*) . . . *Agaricus.*

31 Gills distinctly spotted or distinctly mottled; stipe stiff but breaking with a snap when bent; growing on dung or in richly manured areas. . . . *Panaeolus.*

Gills not spotted nor distinctly mottled; stipe cartilaginous or not; and fruit-body rarely growing on dung. . . . 32

32 Gills broadly attached to the stipe (adnate) and with a veil girdling the stipe. . . . *Stropharia.*

Gills narrowly attached to the stipe (adnexed) or with concave dentation near the stipe (sinuate), or if adnate lacking a ring. . . . 33

33 Gills with concave indentation near the stipe (sinuate) and cap and stipe with a cobweb-like veil. . . . *Hypholoma.*

Gills attached to the stipe but lacking a distinct concave indentation near the stipe. . . . 34

34 Stipe stiff but breaking with a snap when bent; edge of cap incurved at first and cap-surface composed of filamentous cells. . . . *Psilocybe.*

Stipe fragile, edge of cap straight even when young and cap-surface composed of rounded cells. . . . *Psathyrella.*

35 Fruit-body fleshy and readily decaying, often firm but never tough. . . . 36

Fruit-body tough and not easily decaying. . . . 37

36 Growing on other agarics. . . . *Asterophora* (Nyctalis). (and some *Collybia*)

Not growing on other agarics. . . . 37

37 Spore-bearing layer (hymenium) on fold-like often forked gills or simply on irregularities. . . . 38

Spore-bearing layer (hymenium) on distinct well-formed gills. . . . 39

38 Spore-bearing layer on fold-like gills. . . . *Cantharellus.*

Spore-bearing layer on smooth or irregular surface. . . . *Craterellus.*

39 Cap easily separable from the stipe. . . . 40

Cap not easily separable from the stipe. . . . 41

40 Stipe with girdling veil (ring) and/or with a persistent cup-like structure at the base (volva); cap usually with warts or scales distributed on its surface. . . . *Amanita.*

Stipe with ring but lacking the volva; cap surface powdery, hairy or scaly. . . . *Lepiota* & related genera.

41 Cap, stipe and gills brittle; stipe never stiff and either exuding a milk-like juice or not; spores with spines or warts which stain blue-black in solutions containing iodine. . . . 42

Cap, stipe and gills soft or if stipe is stiff then snapping when bent and gills never brittle. . . . 43

42 Fruit-body exuding a milk-like or coloured fluid. . . . *Lactarius.*

Fruit-body not exuding fluid. . . . *Russula.*

43 Gills thick, watery and lustrous (waxy) or with a bloom as if powdery with talc; often brightly coloured. . . . 44

Gills not waxy and rarely over 1·5 mm thick. . . . 46

44 Gills rather watery and lustrous (waxy); spores smooth. . . . 45

Gills rigid not watery, with powdery bloom; spores with distinct spines. ... *Laccaria*.

45 Fruit-body with a distinct veil and growing in woods; cap often viscid or pale coloured. ... *Hygrophorus*.
Fruit-body lacking a veil and usually growing in fields; cap usually brightly coloured and sometimes viscid. ... *Hygrocybe*.

46 Stipe with girdling veil (ring) and/or stipe not attached to the centre of the cap (eccentric). ... 47
Stipe central and lacking a ring. ... 48

47 Stipe central and possessing a ring. ... *Armillaria*.
Stipe not centrally attached to the cap (members of the Pleurotaceae) including *Pleurotus* (Oyster mushroom).

48 Stipe fibrous. ... 49
Stipe stiff only in the outer layers. ... (52)

49 Gills with a concave indentation near the stipe (sinuate). ... 50
Gills attached to and descending down the stipe (decurrent). ... 51

50 Spores with warts which darken in solutions containing iodine. ... *Melanoleuca*.
Spores not so colouring in solutions containing iodine. ... *Tricholoma* & related genera.

51 Spores with warts which darken in solutions containing iodine. ... *Leucopaxillus*.
Spores not so colouring in solutions containing iodine. ... *Tricholoma* & related genera.

52 Gills thick and with rather blunt edges. ... *Cantharellula* & *Hygrophoropsis*.
Gills thin and with distinct sharp edge. ... 53

53 Gills attached to and descending down the stipe (decurrent); cap often depressed at the centre and sterile cells absent from the gills and the surface of the cap. ... *Clitocybe* & *Omphalina*.
Gills attached to the stem but not descending down the stipe (adnate to adnexed), or if descending then distinct sterile cells on the gills, cap and stipe. ... 54

54 Cap-edge straight and usually striate when young; cap thin and somewhat conical and gills descending down the stipe or not. ... *Mycena* & related genera.
Cap-edge incurved; non-striate and cap rather fleshy; gills not descending down the stipe. ... 55

55 Stipe dark and woolly at least in the lower half and the cap viscid; fruit-bodies growing in clusters on tree trunks. ... *Flammulina*.
Stipe not dark and woolly. ... 56

56 Cap viscid and stipe usually rooting; fruit-body growing directly on wood or attached to wood by long strands or cords of mycelium (rhizomorphs). ... *Oudemansiella*.
If cap viscid and fruit-body neither attached to wood by cords of mycelium nor stipe with a rooting base. ... *Collybia* & related genera.

57 Stipe central and gills often interconnected by veins; can be dried and later

revived purely by moistening. ... *Marasmius* & related genera.
Stipe not attached to the centre of the cap and fruit-body, although persistent, not easily revived to natural shape after once being dried. ... 58

58 Spore-print blue-black with solutions containing iodine. ... 59
Spore-print yellowish in solutions of iodine. ... 60

59 Gills toothed or notched along edges. ... *Lentinellus.*
Gills even along their edges and not toothed. ... *Panellus.*

60 Gills appearing as if split down their middle. ... *Schizophyllum.*
Gills not splitting. ... 61

61 Gills notched or toothed along their edges. ... *Lentinus.*
Gills even along their edges and not toothed. ... *Panus.*

62 Spore-print yellowish, purplish-black or pink. ... 63
Spore-print some shade of brown, but without purplish flush. ... (66)

63 Spore-print yellowish or pinkish. ... 64
Spore-print purplish-brown or blackish. ... 65

64 Spore-print yellowish. ... *Gyroporus.*
Spore-print pinkish. ... *Tylopilus.*

65 Spore-print purplish-brown. ... *Porphyrellus.*
Spore-print blackish and spores ornamented. ... *Strobilomyces.*

66 Cap glutinous and stem with or without girdling veil (ring); sterile cells (cystidia) within tubes clustered together. ... *Suillus.*
Cap at most viscid and then only in wet weather and sterile cells within tubes individually sited. ... 67

67 Stipe-surface covered with distinct black or dark brown, or white then darkening scales; spore-print clay-brown with or without a flush of cinnamon-pinkish brown. ... *Leccinum.*
Stipe-surface covered completely or in part with a network or pattern of faint lines, or pale yellow or red-rust but never black dots; spore-print olivaceous-buff. ... *Boletus* & related genera.

68 Spore-bearing layer (hymenium) quite smooth, or spread over veins, or shallow pores; fruit-body top-shaped, fan-shaped or club-shaped, or spread over the substrate (resupinate). ... 69
Spore-bearing layer lining the inner surface of tubes or borne on warts or spines. ... (84)

69 Fruit-body club-shaped, coral-shaped or distinctly funnel-shaped, fan-like or resembling an agaric. ... 70
Fruit-body resupinate or with poorly developed cap. ... (78)

70 Fruit-body coral-like or club-shaped with clubs grouped or branched. ... 71
Fruit-body resembling an agaric or funnel-shaped to fan-shaped. ... (76)

71 Fruit-body large, branched with flattened and curled lobes and so resembling a cauliflower. ... *Sparassis.*
Fruit-body of single or grouped clubs, or if branched then not resembling a cauliflower, the lobes being cylindrical or only slightly flattened and hardly bent. ... 72

72 Fruit-body small arising from a seed-like structure or growing attached to dead herbaceous plant remains. ... 73
Fruit-body medium to large, simple or branched and usually growing on the

ground; one large species grows on wood. ... 74

73 Fruit-body arising from a seed-like body embedded in the plant tissue or found loose in the soil. ... *Typhula*.

Fruit-body on dead plant remains but seed-like structure absent.. . *Pistillaria*.

74 Fruit-body much branched; spores ornamented.

... *Ramaria*. (see also *Thelephora* below).

Fruit-body simple or if with well-developed branches then spores smooth.... 75

75 Fruit-body branched irregularly with many to few branches, grey, white or dull-coloured; spores large, subglobose and smooth. ... *Clavulina*.

Fruit-body club-shaped or if branched then brightly coloured and spores not large and subglobose. ... *Clavaria, Clavulinopsis* & *Clavariadelphus*.

76 Fruit-body resembling an agaric with spores borne on fold-like, often forked and shallow ridges and veins, and often brightly coloured.

... *Cantharellus*. (compare very carefully with *Craterellus* below).

Fruit-body funnel-shaped or fan-shaped. ... 77

77 Fruit-body often dull-coloured or grey with smooth or slightly veined outer surface. ... *Craterellus*.

Fruit-body wrinkled, irregular or smooth and powdery, lilaceous to chocolate-brown in colour.

... *Thelephora*. (in N. America compare carefully with *Polyozellus* which resembles a cluster of irregular funnels and *Craterellus* above).

78 Fruit-body sessile or resupinate and fleshy; spores borne on veins united to form shallow pores. ... 79

Fruit-body resupinate or bracket-like and spore-surface veined or rugulose but lacking distinct pores. ... 80

79 Spores colourless. ... *Merulius*

Spores brown. ... *Serpula*.

80 Spore-bearing layer containing long brown spines. ... *Hymenochaete*.

Fruit-body lacking spines although often having encrusted sterile cells. ... 81

81 Surface of fruit-body more or less radiately veined. ... *Phlebia*.

Surface of fruit-body not radiately veined. ... 82

82 Spores brown. ... *Coniophora*.

Spores colourless. ... 83

83 Flesh distinctly formed and fruit-body with or without a reflexed cap.

... *Stereum* & related genera.

Flesh poorly differentiated and fruit-body lacking a cap.

... Members of the *Corticiaceae* (including *Peniophora* & *Hyphodontia*).

84 Spores borne on teeth or spines. ... 85

Spore-bearing layer lining tubes or elongate pores. ... (89)

85 Fruit-body with central stipe; agaric-like but not attached to cones. ... 86

Fruit-body encrusting or bracket-like or with lateral stipe if resembling an agaric. ... 87

86 Fruit-body fleshy; spores smooth. ... *Hydnum* & related genera.

Fruit-body rubbery or tough; spores rough. ... *Hydnellum* & related genera.

87 Fruit-body growing attached to cones and cap with lateral stipe.

... *Auriscalpium*,

Fruit-body not on cones and distinct stipe lacking. ... 88

88 Spores borne on a series of radially arranged notches resembling gills.
. . . *Lentinellus*.
Spores borne on a resupinate layer of spines. . . . *Mycoacia* & related genera.
89 Tubes free one from another; resembling a piece of flesh or liver. . . *Fistulina*.
Tubes united to form a distinct tissue; resembling wood, leather or cork. . . . 90
90 Fruit-body perennial and exhibiting more than one layer of tubes. . . . 91
Fruit-body annual although it can persist in a dried depauperate form for several months. . . . (94)
91 Spores brown. . . . 92
Spores colourless. . . . 93
92 Large brown cells present in the tubes; spores simple.
. . . *Phellinus* & *Cryptoderma*.
Brown sterile cells absent from tubes; spores complex. . . . *Ganoderma*.
93 Large woody fruit-body with crust-like top. . . . *Fomes*.
Medium-sized to small; fleshy tough fruit-body with downy or crust-like top. . . . *Oxyporus, Fomitopsis* & *Heterbasidion*.
94 Spores borne in labyrinth-like elongate pores, cap either poorly developed or absent, and only resupinate pore-surface present. . . . 95
Spores borne in distinct pores on well-developed woody fruit-bodies. . . . (98)
95 Spores borne in labyrinth-like pores. . . . *Daedalea* & *Daedaleopsis*.
Spores borne in elongate pores like very thick gills, or fruit-body completely resupinate. . . . 96
96 Spore-layer in elongate pores. . . . *Lenzites* (white) & *Gloeophyllum* (brown).
Spore-layer consisting of a resupinate pore-layer. . . . 97
97 Pore-layer totally resupinate; flesh very poorly developed.
. . . *Fibuloporia* & related genera.
Fruit-body resupinate or developing ill-formed caps at the margin; flesh well-developed and quite tough. . . . *Datronia, Gloeoporus* & *Bjerkandera*.
98 Fruit-body with a distinct stipe. . . . 99
Fruit-body sessile or with a poorly developed stipe, or if merely with a basal swelling then pores darkening or bruising on handling. . . . 100
99 Pores dark-coloured but spores pale-coloured in mass.
. . . *Coltricia*. (also see *Phaeolus* below).
Pores white or creamy, foot often darkened black and pores hyaline.
. . . *Polyporus*.
100 Pores brightly coloured, red, lilaceous or orange to apricot colour. . . . 101
Pores never as brightly coloured, cream, white, grey or in some shade of brown. . . . 102
101 Pores red to orange-red. . . . *Pycnoporus*.
Pores lilac to violacaeous, or lilaceous-orange to apricot colour.
. . . *Haplopilus* (orange to apricot).
. . . *Hirschioporus* (lilaceous).
102 Pore-surface brown or dark-grey and spores often colourless. . . . 103
Pore-surface white or creamy, or yellow; spores hyaline. . . . 105
103 Pore-surface firm and grey. . . . *Bjerkandera*.
Pore-surface greenish-yellow, bruising brown or yellow-brown and darkening

with age. ... 104

104 Fruit-body lacking a stem, rust-brown, breaking easily, cheesy in texture and with a silky sheen. ... *Inonotus*.
Fruit-body with a broad basal hump, fibrillose spongy with yellow margin to cap. ... *Phaeolus*.

105 Tubes forming a layer quite distinct from the flesh; fruit-body fleshy and tough. ... 106
Tubes not forming a layer distinct from the flesh; fruit-body woody or corky. ...(110)

106 Pore-surface bright yellow; upper surface yellow or orange. ... *Laetiporus*.
Pore-surface white; upper surface usually dull coloured or white. ... 107

107 Fruit-body medium to large, shell-shaped, whitish-brown or silvery-grey on top; on birch. ... *Piptoporus*.
Fruit-body often frond-like, infrequently shell-shaped and if on birch then small. ... 108

108 Fruit-body fan-shaped or frond-shaped, composed of innumerable more or less complete caps joined together at their base or to half way. ... *Grifola* & *Meripilus*.
Fruit-body neither fan-shaped nor frond-shaped and compound. ... 109

109 Fruit-body wholly pale-coloured, white, cream, ivory etc. ... *Tyromyces*.
Fruit-body except pores usually some shade of brown. ... *Polyporus*.

110 Cap thick corky or woody and pores medium or large. ... *Trametes* & *Pseudotramets*.
Cap thin but leathery and pores small. ... *Coriolus*.

111 Fruit-body club-shaped or coral-like. ... *Calocera*.
Fruit-body top-shaped or with irregular bumps. ... 112

112 Fruit-body top-shaped. ... *Ditiola*.
Fruit-body cushion-like or brain-like or with irregular bumps. ... *Dacrymyces*.

113 Fruit-body lacking a cap and more or less forming a gelatinous coating on plant debris. ... *Helicobasidium*.
Fruit-body with more or less distinct cap; gelatinous but tough. ... 114

114 Fruit-body ear-like or cup-shaped; upper surface with grey hairs and lower surface lilaceous-brown or Burgundy-coloured. ... *Hirneola*.
Fruit-body at first cup-shaped but then spreading; upper surface grey and hairy, and lower surface purplish. ... *Auricularia*.

115 Fruit-body with distinct stipe and spines on lower surface.. ... *Pseudohydnum*.
Fruit-body lacking a well-developed stipe, the latter either reduced to a small lobe or entirely absent. ... 116

116 Fruit-body flattened or disc-shaped, often with warts or veins on the surface; spores more or less sausage-shaped. ... *Exidia*.
Fruit-body brain-like or with irregular bumps, sometimes lobed or irregular and encrusting. ... 117

117 Fruit-body brain-like or with bumps or bosses; spores rounded to ovoid. ... *Tremella*.
Fruit-body encrusting woody or herbaceous material; spores ellipsoid. ... *Sebacina*.

B—*Key to major groups based on characters of fruit-body and spores (only large and obvious genera included)*

118 Asci borne on a distinctly stalked fruit-body. ... 119
Asci borne on an irregularly lobed, rounded, or club-shaped fruit-body, or the latter cup-shaped, but never stalked. ...(123)
119 Cap cup-shaped. ... 120
Cap honeycomb-like, saddle-shaped or irregular. ... 121
120 Fruit-body grey. ... *Cyathipodia* (see also 130).
Fruit-body red within. ... *Sarcoscypha*.
121 Cap irregularly chambered to honeycomb-like.
... *Morcella* and allies (Morels)
Cap saddle-shaped or irregular. ... 122
122 Stipe stout, furrowed, ribbed or chambered. ... *Helvella*.
Stipe slender with even surface. ... *Leptopodia*.
123 Fruit-body growing beneath soil-surface.
... *Tuber, Elaphomyces* and allies. (True & False truffles).
Fruit-body not growing beneath soil-surface. ... 124
124 Fruit-body black and carbonaceous either within, externally or throughout.
... 125
Fruit-body brightly coloured, or if brown then soft and pliable. ...(127)
125 Fruit-body hemispherical with distinct concentric zones of growth when cut. ... *Daldinia*.
Fruit-body variously shaped or if hemispherical then without zonation. ... 126
126 Fruit-body club-shaped, cylindrical or spindle-shaped.
... *Xylosphaera*. (Dead man's fingers: Stag's horn fungus).
Fruit-body hemispherical or cushion-shaped.
... *Ustulina* & *Hypoxylon*. (if growing on pore-fungi see *Hypocrea*).
127 Fruit-body distinctly cup-shaped or ear-shaped. ... 128
Fruit-body irregularly lobed, undulating. ... *Rhizina*.
128 Fruit-body with a split down one side. ... *Otidea*.
Fruit-body cup-shaped or at most with a wavy margin. ... 129
129 Fruit-body orange or red; spores ornamented with ridges and reticulations.
(margin with short brown hairs). ... *Melastiza*.
(margin with eyelash-like hairs). ... *Scutellinia*.
(margin naked like orange peel). ... *Aleuria*.
Fruit-body duller in colour, yellow, brown, violaceous but never orange or red; spores smooth or minutely warted or faintly netted. ... 130
130 Spore-bearing layer becoming bluish-green in solutions containing iodine; (if with stalk then rudimentary). ... *Peziza*.
Spore-bearing layer not blueing with iodine solutions; cup with broad, ribbed or furrowed stalk-like base. ... *Paxina*.

Glossary of Technical Terms

Acidulous: slightly sour.
Adnate (of the gills or tubes): broadly attached to the stipe for at least one quarter of their length.
Adnato-decurrent (of the gills): broadly attached and running down the stipe.
Adnexed (of the gills): narrowly attached to the stipe by less than one quarter of their length.
Adpressed: closely flattened down, as if by a domestic iron. Used to describe the scales on a cap or stipe.
Amygdaloid (of the spores): like an almond.
Amyloid (of the spore walls): turn greyish or bluish or blackish-violet in solutions containing iodine.
Anastomosing: joining together.
Apex (of stipe or fruit-body): the top.
Apical: belonging to the apex.
Apiculus (of the spores): the short peg-like structure at the basal end of the spore by which it is attached to the basidium.
Appendiculate (of gill fungi): where the expanded cap-edge is fringed with tooth-like velar remains.
Arcuate (of the gills): bent as a bow.
Ascus: a sac-like body in which the spores are borne, usually eight. (pl. asci)
Ascomycetes: fungi producing asci (cup fungi and their allies).
Asexual: without sex.
Astringency: the property of drawing together soft tissues, contracting severed vessels, thus checking blood flow.
Attenuated: growing slender towards the extremity.
Auto-digesting: becoming fluid when mature, as with many coprinus.

Basidium: a club-shaped body on which spores are borne externally on stalks. (pl. basidia)
Basidiomycetes: fungi bearing basidia.
Bifid: divided halfway into two.

Caespitose (of the fruit-body): growing in tufts.
Campanulate: bell-shaped.
Carbonaceous: hard and containing very dark pigment.
Cartilaginous: hard and tough.
Chlamydospores: asexual spores with a very thick wall.
Cinereous: having the colour of wood-ashes.
Clavate (of the stipe): club-shaped.
Concolorous: of the same colour.
Confluent: united at some point.
Conidium: as asexual spore. (pl. conidia)
Convex: swelling on the exterior surface into a rounded form.
Coralloid: branching like coral.
Cortina (of agarics): the cobweb-like veil, in many young toadstools, between cap margin and stipe.
Corymbose: like a corymb in which the stalks of the lower flowers are longer than those of the upper so that they rise to the same level.
Crenate: where an edge is indented, scalloped or has rounded notches.
Crispate: having a crisped, curled appearance.
Crustaceous: hard and brittle, having a crust.
Cuspidate: having a sharp end or point.
Cuticle (of cap or stipe): the outer skin.
Cystidium: a sterile differentiated end-cell usually on the surface and edges of the cap, gill and stipe. (pl. cystidia)

Decurrent (of the gills and tubes): with a part attached to and descending down the stipe.
Deliquescing: see auto-digesting.
Dentate: toothed.
Denticulate: with little teeth.
Dextrinoid (of spores): when spore mass turns red-brown in Melzer's iodine.
Dichotomous: forked into two equal branches.

Eccentric (of the cap): laterally placed on the stipe.
Echinate: having sharp points or set with prickles like a hedgehog.
Echinulate: with short bristles.
Effused: irregularly spread over substrate.
Ellipsoid (of the spores): elliptic in outline in all planes.
Emarginate (of the gills): with a sudden very conspicuous curve as if scooped out at point of attachment to stipe.
Endoperidium: inner layer of limiting membrane of a fruit-body (Gastromycete) e.g. Puff-Balls and Earth-Stars.
Esculent: something that is edible.
Evanescent: liable to vanish.
Exoperidium: outer layer of limiting membrane of a fruit-body (Gastromycete) e.g. Puff-Balls and Earth-Stars.

Ferruginous: the colour of iron rust.
Fetid or Foetid: having an offensive smell.
Fibrillar: see fibrillose.
Fibrillose (of the cap and stipe surfaces): clothed with small fibres.
Fibrilloso-striate: clothed with small fibres and marked with fine lines.
Fissure: a cleft, a deep narrow depression.

Flaccid: flabby, loose or limp.
Flexuose: see flexuous.
Flexuous: winding, wavering or curving.
Floccose: having a loose cottony surface.
Flocci: cotton-like tufts.
Floccules: small tufts.
Frondose: deciduous or broad-leaved trees.
Fructification: the act or process of becoming fruitful.
Fruit-body: the whole agaric, toadstool or mushroom.
Fugacious: soon disappearing.
Fuliginous: sooty, smoky, dusky.
Furfuraceous: mealy or scurfy.
Fuscous: brownish-black.
Fusiform: spindle-shaped, tapering at both ends (two dimensions).
Fusoid: spindle-shaped (three dimensions).

Germ-pore: a differentiated apical and usually thin-walled portion of the spore.
Gibbous: hump-backed.
Glabrous: free of hair, smooth.
Glaucous: of a sea-green colour.
Gleba: the spore-bearing tissue especially of Gastero-mycetes and Tuberales.
Glutinous (of the cap or stipe): having a sticky jelly-like coating.
Gregarious: numerous specimens growing closely together.

Hispid: with short stiff hairs.
Homogeneous: of the same kind or nature.
Hyaline: resembling clear glass.
Hygrophanous (of the cap): translucent when wet, opaque and often paler on drying.
Hymenium: the fertile layer of a fruit-body.

Imbricate: overlapping like tiles on a roof.
Inferior (of ring): low on stipe.
Infundibuliform: funnel-shaped.
Involute: having the edge turned under.

Lacunose (of a surface): being furrowed or pitted, having sunken gaps.
Lanceolate: gradually tapering toward the upper extremity.
Larvae: the growing stage in the life of insects, i.e. caterpillars, maggots.
Lax (of ring): loose, flabby slack.
Linear: like a line, slender.
Lobate: having lobes.

Macroscopic: visible without magnification.
Margin (of cap): the edge or border.
Median (of ring): centrally positioned on stipe.
Mucilaginous: gummy or slimy.
Mycelium: a mass of fungus filaments.
Mycophagist: one who eats fungi.
Mycorrhiza: a symbiotic association of a fungus and the roots of a higher plant.

Opaque: not transparent.
Ovate: egg-shaped (two dimensions).
Ovoid: egg-shaped (three dimensions).

Papilla: a small nipple.
Papillate (of the cap): with a small central nipple-like process.
Partial veil: the veil extending from the cap margin to the stipe.
Peduncle: a stalk.
Pellicle (of gill fungi): detachable, skin-like cuticle of the cap.
Peridioles: rounded bodies which contain the spores in Bird's Nest fungi.
Peridium: the wall or limiting membrane of a fruit-body of Gastromycete.
Perithecia: the flask-shaped structure in Pyrenomycetes which contain the asci.
Pileate: having the form of a cap.
Pileus: the cap.
Plane (of the cap): an even or level surface.
Plasmodium: a mass of protoplasm in the Myxomycetes.
Plicate: folded like a fan.
Pruinose: finely powdered.
Pubescent: with short, soft hairs.
Pulverulent: dusty, consisting of fine powder.
Punctate: having a pin-pricked surface.
Pungent: affecting the organs of smell or taste with a prickly sensation.
Pyriform: pear-shaped.

Resupinate (of fruit-bodies): lying flat on the substratum with the spore-bearing tissue facing outward.
Reticulate: in the form of a net.
Rhizoids: root-like structures.
Rimose: where hyphae of the cap become slightly separated radially showing underlying tissue.
Rivulose: marked with lines like rivulets.
Rubiginous: red-coloured.
Rugose: roughly wrinkled.
Rugulose: minutely wrinkled.

Saprophytic: living on decaying vegetable matter.
Sclerotium: a hard and often round mass of closely packed hyphae.
Seceding: withdrawing or becoming cut-off voluntarily.
Septate: having a division or cell-wall.
Serrate: edged with teeth as a saw.
Serrulate: furnished with very minute teeth or notches.
Sessile: without a stipe.
Setae: small bristles.
Sinuate (of the gills): having a concave indentation of

that part of the edge nearest the stipe.
Spermatic: of a distinctive, rank, unpleasant, strong, 'earthy' odour.
Spinulose: covered with spines.
Sporangium: an organ containing an indefinite number of spores. (pl. sporangia)
Sporophore: fruit-body.
Squamules: small scales.
Squarrose (of scales): turned back at right angles.
Stellate: resembling a star.
Stipe: stem.
Stipitate: supported by a stipe.
Striae: thread-like.
Striate: having minute lines.
Strigose: beset with long firm hairs.
Subdistant (of gills): between crowded and fully distant.
Subglobose: almost globular in form.
Substrate: the underlying medium from which a fruit-body develops.
Sulcate: grooved or furrowed.
Superior (of the ring): positioned high-up on the stipe.

Tenacious: holding fast, adhesive.
Terrestrial: found growing on the ground.
Tomentose: densely matted and woolly.
Tortuous: twisted.
Translucent: nearly transparent.
Tremulous: trembling or quivering.
Tuberculate: having small wart-like pimples that are usually visible to the naked eye.
Turbinate: top-shaped.

Umbilicate: having a small localised depression.
Umbilicato-depressed: having a depressed centre.
Umbo: a broad central swelling like the boss on a shield.
Umbonate (of cap): having an umbo.
Undulate: wavy.
Uniseriate: arranged in a single series.
Universal veil: the outer envelope in agarics which entirely covers the immature fruit-body.

Velar: pertaining to the universal or partial veil.
Ventricose (of the gills): protruding at the middle.
Vernal: appearing in the spring.
Villous: covered with long fine hairs.
Vinaceous: the colour of wine.
Viscid (of the cap or stipe): very slippery to the touch, sticky or slimy.
Volva: a persistent cup-like structure at the base of the stipe.

*The numbers **1–252** correspond with the illustrations*

SMALLER FUNGI OF ECONOMIC IMPORTANCE

1 Ceratocystis ulmi (Buisman) C. Moreau
Dutch elm disease

Our illustration shows several mature Common Elms which have been killed by this fungal infection.

In recent years, the progression of this disease in England and Wales has been astounding and skeleton elms are at this moment (1977) part of our countryside scene. Needless to say, it also occurs in North America, probably arriving there in unbarked timber imported from Europe. The disease is carried from tree to tree by bark beetles of the genus *Scolytus* and by means of root grafts from diseased to healthy trees; there may also be other carriers. The fungus affects the trees by indirectly blocking up their water-conducting vessels and also by direct poisoning.

A great amount of work is being and has been carried out, both in Europe and America, in an effort to find a chemical control for the disease, including systemic insecticides and fungicides.

2 Leocarpus fragilis (Dicks.) Rost.
Brittle smooth-fruit

Our illustration shows the plasmodial state.

A common Myxomycete or Slime-mould (not a fungus in the true sense). Found in and on decaying branches and stumps during summer and autumn. There is no mycelium produced; the lemon then orange-lemon plasmodium grows and expands in the substrate by feeding on bacteria and finally produces the characteristic fruit-bodies or sporangia. These are visible on the surface as clustered obovoid or globose blobs, sessile or stalked. Yellowish-brown, then chestnut or bay-brown, smooth and from 2 to 4 mm. long.

The spores are spinulose and purplish-brown or dark brown. 9–13 μm in diameter.

3 Rhytisma acerinum (St. Amans) Fries
Rhytisma pseudoplatani
Tar spot fungus or Maple blotch

Black blotches which occur on the leaves of sycamore and maple. They are slightly raised above the natural surface of the leaf (lens). Yellow at first but soon becoming black.

1 Ceretocystis ulmi

It is parasitic and weakening to the host. A common species which begins to appear in early summer.

4 Puccinia coronata Cd.
Coronated rust

One of the many rusts which are instrumental in causing failure and ill-health in both wild and cultivated vascular plants. When affected leaves are examined under a microscope, they will be found to be covered by discreet packets of orange-brown spores; the mycelial threads are found only inside the leaves. In maturity, the spores drop down and many lie in masses on the leaves, others are blown or carried to neighbouring plants. The mycelium penetrates the cells by minute suckers, thereby absorbing nutrient and causing ill-health to the host.

Coronated rust is not uncommon on various grasses in autumn. Our illustration features it on Yorkshire Fog (*Holcus lanatus*).

5 Ustilago hypodytes Fries
Grass-culm smut

The Smuts are another group of micro-fungi which can prove troublesome to the commercial grower, especially in gramineous plants. Grass-culm smut first appears on the culms beneath the sheaths and later becomes exposed as a dusty dirty-brown covering. Often it can be found on various species of grass during summer.

Our illustration shows it on Sea-lyme grass (*Elymus arenarius*).

2 Leocarpus fragilis

3 Rhytisma acerinum

4 Puccinia coronata

5 Ustilago hypodytes

ASCOMYCOTINA—CUP FUNGI and ALLIES

Order *PEZIZALES*

The fruit-body is very small to medium sized, globe-shaped, then expanding until cup-like or disc-shaped, sometimes even plane or convex. Sessile or somewhat stipitate. The larger species are mainly terrestrial. The hymenium is on the upper or inner surface of the cup.

Most described species are edible but unsubstantial.

6 Aleuria aurantia (Fries) Fuckel
Peziza aurantia Fries
Orange-peel fungus

At first resembles a small pale pink sphere but after increasing in size it breaks open in the form of an irregular cup of 3–12 cm across. Orange-red inside and rosy or pale orange outside.
Stipe: Absent. This fungus is sessile.
Spores: 18–22 × 9–11 μm reticulated.
Flesh: White, thin and fragile, with an agreeable odour.
Habitat and season: Likes sandy soil but can be found on bare soil in woods, also on paths and lawns during late summer and autumn. A common species sometimes growing in large groups or clusters.
Edibility: May be eaten raw in salads.

* **Neotiella rutilans** (Fries) Dennis
Peziza rutilans Fries

Similar in shape and colour to *A. aurantia* but has a short stipe. The cup, which is hairy on the outside, is only 0·5–1·5 cm across.
Spores: oval, reticulated 22–26 × 11–14 μm.
Habitat and season: Can be found on heathland from October to January.
Edibility: No.

* **Peziza violacea** Persoon

Fruit-body: 1–3 cm across. Closed and almost globose at first then shallowly cup-shaped, finally expanded with wavy margin and following the contour of the substrate. Subsessile, fleshy and brittle. When young, the disc is dark violet in entirety, later umber to purplish-brown and finally drying out violet with brownish tone. The outer surface is greyish to light brown pruinose.
Spores: Smooth and cylindrical. 13–15 × 7–9 μm. Asci 8-spored.
Habitat and season: Uncommon and gregarious on burnt ground or on charcoal. It is also said to occur on rotten tree trunks.
Edibility: Not known.

* **Peziza repanda** Persoon
Discina repanda Saccardo

Fruit-body: 2–10 cm across. Shallowly cup-shaped then expanded with margin split and wavy and somewhat reflexed. The base is infrequently stem-like and rooting. The disc is pale or dark chestnut to hazel brown. The outside is whitish to pale fawn and finely mealy.
Spores: Smooth, very long and elliptical. 15–16 × 9–10 μm. Asci 8-spored and cylindrical.
Habitat and season: Common in clusters or scattered on bare ground in woods especially beech; on decayed tree trunks and on sawdust. Also in farm yards.
Edibility: Unknown.

7 Otidea onotica (Fries) Fuckel
Peziza onotica Pers. ex Fries
Lemon-peel fungus

Yellowish and shaped like a donkey's ear. It grows 3–9 cm high with lobes up to 4 cm across. Often growing in clusters.
Spores: Elliptic, 10–14 × 5–7 μm.
Stipe: A short peduncle with white hairs.
Habitat and season: Grows under frondose trees and can occasionally be found in August and September.
Edibility: Is said to be edible.

* **Otidea leporina** (Fries) Fuckel
Peziza leporina Batsch ex Fries

Of a similar shape to *O. onotica* but somewhat smaller. This tan-coloured fungus grows on spruce needles and may be found occasionally from August to October.
Spores: Elliptic and smooth, 12–14 × 7–8 μm.

8 Melastiza chateri (W. G. Smith) Boudier

Cup: 0·5–1·5 cm across. Closed at first, then expanded. Margin erect or slightly incurved, minutely fringed. Disc deep orange or crimson inside, pale reddish-brown outside; sessile. Usually irregular in form through mutual pressure.
Spores: Elliptical and hyaline. 13–16 × 7–8 μm. Asci 8-spored.
Habitat and season: Grows densely caespitose on woodland paths and bare ground etc. Common during late summer and autumn. A striking and beautiful species especially when covering large areas.
Edibility: Not known.

6 Aleuria aurantia

7 Otidea onotica

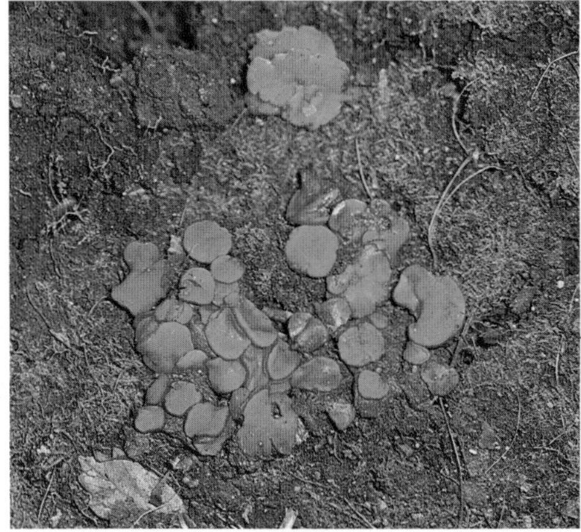

8 Melastiza chateri

9 Sarcoscypha coccinea Saccardo
Peziza coccinea Scop. ex Fries
Scarlet elf-cup

Cup: 2–6 cm in diameter and finally expanded. Disc deep rose-red or carmine, exterior whitish-grey or pinkish and tomentose.
Spores: Smooth and fusiform. 26–34 × 11–13 μm.
Stipe: Varies in length according to the position of the fruit-body, and is often curved. The base is attached to the substrate by a mass of whitish mycelium.
Habitat and season: Locally common. Gregarious on dead branches (especially hazel) lying on the ground in woods. From December to April. It is a beautiful fungus and often used in the home for decorative purposes.
Edibility: Yes.

There is a variety: 'Albida' which has the disc cream-coloured, but otherwise conforms with the type form.

10 Otidea bufonia (Persoon) Boudier.
Peziza bufonia Persoon

Cup: 3–7 cm. Globose at first and closed, then hemispherical and fragile. Narrowed into a short stem-like base which is often rooting. Cup margin entire or notched. Inside of cup (disc) bright brown. Outside of cup a similar colour or maybe a little duller with violaceous tints and warty.
Spores: Hyaline, smooth, elliptical and oblong with rounded ends. 15–20 × 10–12 μm. Asci 8-spored.
Habitat and season: Common on the ground in moist woods and on rubbish tips etc. From August to April. Not recorded in America.
Edibility: Worthless.

* **Peziza vesiculosa** Bulliard.

Cup: 3–7 cm. Globose at first and closed, then hemispherical. Margin usually remaining incurved and notched. Disc pale brown. Outside of cup brownish and coarsely granular from the presence of minute and irregular warts.
Spores: Hyaline, smooth and elliptical, 21–24 × 11–12 μm. Asci 8-spored.
Habitat and season: Grows clustered and distorted on rich soil and manure heaps etc. August to April.
Edibility: Worthless.

* **Otidea cochleata** ([Linn.] St Amans) Fuckel
Peziza cochleata [Linn.] St Amans

Cup: 5–8 cm. An uncommon species that bears similarity to *O. bufonia* but outside of the cup is primrose.
Spores: Hyaline and smooth, 16–18 × 7–8 μm.
Habitat and season: Usually in grass. Occasionally found July to April.
Edibility: Worthless.

11 Peziza badia Mérat

Cup: 3–5 cm. Subglobose and closed at first, then hemispherical or more expanded. Margin more or less entire. The whole fruit-body is rather thick and often wavy. Inside of cup dark brown. Outside of cup paler brown and minutely granular, often with a purple tinge.
Spores: Hyaline and elliptical with one large oil drop, irregularly reticulate. 17–20 × 9–12 μm. Asci 8-spored.
Habitat and season: Common on the ground, usually gregarious. Found in grass and bare, scorched or burnt places. August to November.
Edibility: Worthless.

* **Peziza succosa** Berkeley

Cup: 1·5–2·5 cm. Hemispherical with incurved margin at first. Inside of cup (disc) pale yellowish-brown with olive or violet tinge. Outside of cup paler or greyish-brown and minutely scurfy. Broken flesh exudes a liquid which turns yellow.
Spores: Hyaline and elliptical, with warts and ridges. 18–22 × 10–12 μm.
Habitat and season: Grows on bare soil in shady places. From August to October.
Edibility: No.

12 Disciotis venosa (Persoon) Boudier POISONOUS

Cup: 3–10 cm across. Margin incurved when young, later expanded becoming split, or lobed and wavy. Sessile or with a short and ample stem-like base. The disc is umber-brown with corrugations at centre, the outside is dirty white and furfuraceous with vein-like ribs radiating from base.
Spores: Hyaline and elliptical, often with one large oil drop. 18–24 × 11–13 μm. Asci 8-spored.
Habitat and season: Fairly common on paths and base ground etc. in spring.
Edibility: *Poisonous.* It has a strong nitrous odour which is characteristic.

9 Sarcoscypha coccinea

10 Otidea bufonia

11 Peziza badia

12 Disciotis venosa

13 Rhizina undulata Fries
Rhizina inflata (Schaeffer) Karsten
Pine fire fungus or Doughnut fungus

Fruit-body: 3–12 cm across and 2–8 cm high. Crust-like and convex. More or less cushion-shaped, irregularly undulated and often lobed. Dark brown with extreme margin much paler and often slightly raised. The under surface is paler and concave, conforming more or less in outline with the upper surface, but furnished with numerous stout whitish rhizoids which are attached to the substrate.
Spores: 24–36 × 8–10 μm.
Habitat and season: Fairly common under conifers, especially in burnt areas. From June to October.
Edibility: No.

Genus-Helvella False Morels

The fruit-body has a lacunose stipe and cap, the latter being supported at the centre and is irregular with drooping lobes; it is saddle-shaped or cup-shaped. The hymenium is borne on the upper surface.

All species are considered edible but should never be eaten raw or in large quantities.

14 Helvella lacunosa Fries
The black helvella, Slate-grey helvella or Elfin saddle

Cap: 2–4 cm. Inflated and lobed. Similar in shape and form to *H. crispa* but the upper surface is smoky-black in colour and the under surface ashy-grey.
Spores: Elliptic. 17–21 × 10–13 μm.
Stipe: Irregular, netted and with longitudinal ridges. At first stuffed with fibres, later hollow with thin walls. White to greyish-black.
Flesh: Thin and tough, White to grey.
Habitat and season: Fairly common on the ground under frondose trees or on hedgebanks, footpaths and verges etc. Usually gregarious. Found in late summer and autumn.
Edibility: Edible when cooked but rather indigestible. Should not be eaten raw.

* **Helvella elastica** Bulliard

Smaller than *H. lacunosa.*
Cap: Has a yellowish upper surface and a lighter under surface.
Stipe: Long, slender and hollow. White to greyish-brown.

* **Helvella monachella** Fries

Also smaller than *H. lacunosa.*
Cap: Has a blackish-violet or brown upper surface and a whitish under surface.
Stipe: Slender and hollow. Whitish and fairly smooth.
Habitat: Not recorded in America.
Edibility: No.

15 Helvella crispa Fries
White helvella

Cap: 2–5 cm high. 3–5 cm across. Thin, two-lobed and saddle-shaped. At first the margin adheres to stipe, later it moves away and curls in an unpredictable manner. The upper surface is smooth and very wrinkled in varying shades of off-white. The underside is pale tan.
Spores: These are borne on the upper side of the lobes. They are long and elliptic. 17–21 × 10–13 μm.
Stipe: 3–6 cm. Very irregular, hollow and often swollen at the base. The prominent longitudinal furrows can give the appearance of a network pattern. Whitish in colour.
Flesh: Thin, tough and white with a pleasant odour.
Habitat and season: Common in woodlands, usually in grassy, sheltered and damp places. Solitary or a few together. May be found from March to April and again August to November.
Edibility: Should not be eaten raw, but is a good esculent when cooked. Young specimens should be chosen.

* **Helvella acetabulum** (Linn. ex St. Amans) Quélet POISONOUS
Paxina acetabulum (Linn. ex St. Amans) O. Kuntze

Fruit-body: Cup-shaped and stipitate. 3–6 cm across and 3–10 cm high. Fleshy and rather tough. Margin involute and often lobed. Disc is dark umber-brown inside, paler and minutely scurfy outside.
Spores: Hyaline, smooth and broadly elliptical with large oil drop. 18–22 × 12–14 μm. Asci 8-spored.
Stipe: Tall and thick, somewhat hollow, with vein-like or parallel ribs which continue upwards for some distance. Whitish.
Habitat and season: Fairly common on paths and bare open ground etc. In the spring.
Edibility: *Poisonous.*

* **Helvella leucomelaena** (Persoon) Nann POISONOUS
Paxina leucomelas (Persoon) O. Kuntze

Similar to *H. acetabulum* but with a blackish-grey disc which is whitish and slightly rough on the outside.
Spores: Hyaline. 22–24 × 12 μm.
Stipe: The ribs do not ascend up the fruit-body.
Edibility: *Poisonous.*

13 **Rhizina undulata**

14 **Helvella lacunosa**

15 **Helvella crispa**

16 **Gyromitra infula**

Genus-*Gyromitra* Brain-fungi

Like Morchella, but the cap has brain-like ribs or folds. Although often referred to as edible, we recommend they be considered and treated as poisonous.

16 Gyromitra infula (Schaeff. ex Fries) Quélet
Physomitra infula (Persoon) Boudier

Fruit-body: 8–16 cm high.
Cap: 4–8 cm broad. Hooded with two to four irregular sulcate, drooping lobes which adhere to the sides of the stipe. Deep reddish-brown or cinnamon; whitish and downy underneath.
Spores: Hyaline, smooth and elliptical, ends obtuse. 21–23 × 11–12 μm. Asci 8-spored.
Stipe: Usually smooth, sometimes compressed. Whitish or tinged reddish. Pruinose and rugose, becoming hollow with age.
Flesh: Thin, fragile and white.
Habitat and season: Occasional on the ground in woods. Autumn and Spring.
Edibility: No.

Genus-Morchella The Morels

The fruit-body has a hollow stipe and a club-shaped, globose or conical cap which is angularly pitted. The hymenium is borne in the hollows, the ribs being sterile.

All are edible and sought after as esculents. They occur in spring.

17 Morchella esculenta St. Amans
Common morel

Cap: 4–8 cm high and 4–6 cm in diameter, the size is very variable. Roughly oval in shape with rather large, irregular, rounded or angular pits which are framed by sinuous ridges. The pits are ochraceous and the ribs more yellowish, sometimes with a pink or olivaceous flush. The base of the cap is adnate to the stipe.
Spores: Creamy in the mass, smooth and elliptic, 16–23 × 11–14 μm. The cylindrical asci are 8-spored.
Stipe: 3–6 cm high. Stout, cylindrical and hollow but not regular and often larger at the base. Rugose and longitudinally grooved. Minutely furfuraceous above. Dirty white in colour, brownish with age.
Flesh: Waxy, brittle and ochraceous. Has a pleasant taste and odour.
Habitat and season: Occasional on the ground in wood clearings, under hedgerows, on banks and burnt ground etc. Found in the spring.
Edibility: Yes, and excellent when cooked slowly. It should be thoroughly cleaned before cooking and large quantities should not be eaten at a sitting. Can also be preserved by drying on strings, taking care that individuals are not touching; this prevents moulds developing.

18 Morchella elata Fries

Cap: 5–7 (25) cm high and almost as broad, variable in size. Conic-ovate, the base being separated from the stipe by a collar. The ribs are strongly elevated and more or less parallel from top to bottom; they are connected by slender cross-bars which are shallower than the main ribs. Olivaceous-brown with paler ribs which darken to blackish.
Spores: Creamy in the mass. 19–25 × 14–15 μm. Asci large, cylindrical and 8-spored.
Stipe: Stout and grooved but hollow and fragile. White to ochraceous.
Flesh: Thin, waxy and brittle. Has a pleasant taste and odour.
Habitat and season: Occasional in pine wood clearings and burnt areas, especially on sandy soil. Found in spring.
Edibility: As *Morchella esculenta* (illustration 17).

* **Verpa conica** Swartz ex Persoon

Cap: 1·5–4 cm high. Campanulate or subglobose and fairly even. Smooth or wrinkled, not reticulated, more or less closely pressed to the stipe but always free from it. The margin usually inflexed. Olive to dark brown above, underside yellowish. The hymenium covers the outer surface of the cap.
Spores: Hyaline or slightly yellowish, smooth and elliptic. 20–24 × 12–14 μm. Asci 8-spored and cylindrical.
Stipe: 3–8 cm in length, cylindric, stuffed then hollow. Paler than cap with irregular transverse reddish-yellow belts composed of slightly darker granules.
Habitat and season: Rare. On the ground on heaths, amongst grass and along roadsides. April to May.
Edibility: Not known.

Order-HELOTIALES

The fruit-body is sessile or stipitate. The disc is plane or convex, waxy and naked.

19 Bulgaria inquinans Fries
Black stud fungus

Fruit-body: Globose at first and blackish-brown. 2–4 cm. Consistency of india-rubber, filled with jelly-like substance. Later opening cup-like, the inside of which is blackish and smooth, the outside is dark brown and scurfy.
Spores: Dark brown or black and colouring substrate, 10·5–14 × 5·5–6·5 μm.
Habitat and season: Common and gregarious on the bark of felled oak and beech. September to November.
Edibility: Worthless.

20 Chlorosplenium aeruginosum (Fries) de Notaris

Fruit-body: 0·5–1 cm in diameter. Top-shaped at first and closed, then expanding cup-shaped, soon flattened. The margin is usually wavy and irregular, flexible and glabrous. All parts are a deep verdigris green, the disc sometimes slightly paler.
Spores: Hyaline or greenish, narrowly cylindric, fusiform, straight or curved. 10–14 × 2·5–3·5 μm. Asci 8-spored.
Stipe: 1–3 mm long, expanding into the ascophore.
Habitat and season: Fairly common on fallen branches of frondose trees especially oak and ash. It stains the timber on which it grows a deep verdigris green. This is known as 'green oak', formerly used and maybe still is, in the furnishing industry.
Edibility: No.

17 Morchella esculenta

18 Morchella elata

19 Bulgaria inquinans

20 Chlorosplenium aeruginosum

21 Leotia lubrica Persoon
The 'gum-drop' fungus

Fruit-body: Nail-shaped and 1–6 cm high. The head is gelatinous and slimy, uneven and involute. 1·5–2 cm across, olive or yellowish-brown.
Spores: Hyaline, smooth and narrowly elliptical, straight or slightly curved. 22–25 × 5–6 μm. Asci 8-spored.
Stipe: 3–5 cm high and more or less equal. Slimy and covered with minute white squamules. Stuffed, then hollow. Yellowish or amber.
Habitat and season: Tufted or gregarious on the ground in woods. Especially in wet places. Common from August to October.
Edibility: Worthless

* **Leotia chlorocephala** Schweinitz

Similar in shape to *L. lubrica* but has a dark green head and a green stipe which is often twisted.
Spores: Smooth and hyaline. 17–20 × 5 μm.

* **Trichoglossum hirsutum** (Persoon ex Fries) Boudier
Geoglossum hirsutum Persoon ex Fries
Hairy earth tongue

Fruit-body: 5–8 cm high and roughly tongue-shaped. Dull black or sooty brown. Irregular, flattened and longitudinally wrinkled. Covered with minute dense hairs that are longest in the broader fertile portion of the fungus.
Spores: Olive-brown, smooth, linear-fusiform. Slightly curved and slightly thicker at the apical half. 110–150 × 6–8 μm multi-septate. Septate paraphyses and setae are present.
Habitat and season: Common in grass from August to October. Solitary or gregarious.
Edibility: Worthless.

* **Geoglossum fallax** Durand

Similar to *T. hirsutum* but with a smooth stipe or minutely squamulose above.
Spores: Brown and smooth. 80–100 μm, to 12 septate. Paraphyses are abruptly swollen above.

* **Microglossum viride** (Fries) Gillet

Stands 2–6 cm high. Olive-green in colour with a mealy stipe.
Spores: Smooth and brown. 15–20 × 5 μm, always hyaline. Finally three septate.
Habitat and season: Common from September to November. Grows in frondose woods.
Edibility: No.

Order-SPHAERIALES

Adult fruit-bodies are hard, darkly coloured and carbonaceous, only occasionally occurring in finger-like minute masses.

Genus-Hypomyces

They are parasitic on Basidiomycetes and take the form of a powdery, cottony and often brightly coloured layer on the gills or pores of the host. In some cases, it spreads to other parts, permeating the entire fruit-body.

22 Hypomyces lactifluorum (Schweinitz ex Fries) Tulasne

Locally frequent on Russulas and Lactarius in N. America. Bright orange, then orange-red or dark red and covering the parasitized fruit-bodies which are considered edible in this condition. Obviously one must be certain of the identity of the host.

Our illustration shows *Russula brevipes* as the host.

* **Hypomyces ochraceus** (Persoon) Tulasne

Locally frequent on several Russulas and Lactarius. Peachy or flesh-coloured forming a web over the gills.

* **Hypomyces aurantius** (Persoon) Tulasne

Found on Polypores. Bright orange often with associated white powdery web. Asexual stage.

Closely related are the following species:

* **Apiocrea chrysosperma** (Tulasne) Syd.

Very common on Boletes, reducing them to a yellow powdery mass.

* **Byssonectria viridis** (A & S) Petch and **Byssonectria lateritia** (Fr) Petch

Both grow on Lactarius, reducing the gills to a contorted mass. Green in *B. viridis* and reddish-chrome in *B. lateritia*. The latter is very common on *Lactarius deliciosus* and *L. torminosus*.

23 Xylosphaera hypoxylon Dumortier
Xylaria hypoxylon (Linn. ex Fries) Greville
Candle snuff fungus or Stag's horn fungus

Fruit-body: Shaped rather like a deer's antler and when young the tips are powdered white with conidia. It grows from 2–7 cm high. On maturity, the protruding perithecia give the tips of the fungus a blackish appearance.
Spores: Bean-shaped, 10–14 × 4–6 μm.
Stipe: Black and felty.
Flesh: White and corky.
Habitat and season: Can be found all the year round growing quite commonly on the dead stumps of deciduous trees.
Edibility: Of no value.
Both *X. hypoxylon* and *X. polymorpha* form long growths if kept in darkness and, attracted as they are towards a light source, can be made to meander about simply by a

21 Leotia lubrica

22 Hypomyces lactifluorum

23 Xylosphaera polymorpha and X. hypoxylon

24 Xylosphaera polymorpha (in cross section)

periodic change of the light source position. It is also said by some that they have luminous mycelia.

24 Xylosphaera polymorpha ([Linn.] Fries) Dumortier
Xylaria polymorpha (Mérat) Greville
Dead man's fingers or Wood club fungus

Fruit-body: This fungus has numerous club-shaped sporophores which when young are covered with pale brown spores—the conidia. In the fertile part, which is swollen, the perithecial tips are very prominent, black and roughly wrinkled. It grows from 3–8 cm high.

Spores: Black in mass, fusiform, 18–30 × 5–8 μm.
Stipe: Black, short and tapering towards the base.
Flesh: White and very tough.
Habitat and season: Found commonly throughout the year growing, usually in clusters, on stumps and logs of deciduous trees. Especially beech.
Edibility: Worthless.
The cross-section through a mature specimen shows the perithecia (spore-producing cavities). This fungus causes root-rot in many trees, particularly apple. Often grows in close proximity to *X. hypoxylon*.

25 & 26 Daldinia concentrica (Fries) Cesati & de Notaris
Cramp ball or King Alfred's cakes

Fruit-body: 2–9 cm in diameter. Hemispherical, even or uneven and hard. At first brown, then dull silvery-black and minutely papillate over entire surface. When split or sawn in two, the internally zoned flesh can be seen. These zones are silvery greyish-black and silky to the touch (see illustration).
Spores: Black, lanceolate and flat on one side. 11–17 × 6–8 μm.
Habitat and season: Common and gregarious on living or dead mature trees. Nearly always on ash, but occasionally on others such as beech and birch etc. Any time of the year.
Edibility: No.
In days gone by, people carried these fruit-bodies on their persons as a charm against cramp; hence the common name.

27 Hypoxylon fragiforme (Fries) Kickx
Hypoxylon coccineum Bulliard ex Fries

Fruit-body: 0·5–1·5 cm in diameter. Subglobose or more or less effused and crustaceous. Finely papillate with perithecial tips. Pink to brick-red becoming brown to black.
Spores: Black and uniseriate, 11·5–15·5 × 5–6·5 μm.
Flesh: Hard and brownish-black.
Habitat and season: Common and very gregarious on the bark of dead or felled deciduous trees, especially beech. July to March.
Edibility: Worthless.

Order-HYPOCREALES

Soft and mainly brightly coloured fruit-bodies.

28 Nectria cinnabarina Fries
Coral spot fungus

The gelatinous pink pads often seen on dead sticks are the asexual stage (*Tubercularia vulgaris*). It is not until later in the year that the flask-shaped fruit-bodies of the perfect state are to be found. They are dark cinnabar in colour and as the perithecia develop so their surface becomes warty in appearance.
Spores: In two rows. 14–20 × 5–7 μm.
Habitat and season: It grows in dense clusters bursting through the bark of damp twigs and branches etc. Especially on the recently felled branches of sycamore. Common throughout the year.
Edibility: No.
Often mistaken for a species of Dacrymyces.

25 Daldinia concentrica

26 Daldinia concentrica (in cross section)

27 Hypoxylon fragiforme

28 Nectria cinnabarina

Order-*CLAVICIPITALES*

Tongue-shaped or drum stick-shaped fruit-bodies which are parasitic on insects, false truffles or grasses.

29 Cordyceps militaris (St. Amans) Link
The Scarlet caterpillar fungus

A parasitic fungus which appears above the ground as a slender reddish-orange club, apparently just growing out of the soil and some 2–4 cm in height. This upper fertile portion has a slightly roughened surface caused by the many openings to the perithecia. If the downward tapering club is traced into the soil it will be found to be growing from the dead larva or pupa of a moth or butterfly and usually issues from directly behind the head of its host. When cut open, it will be seen that the contents of the insect have been replaced by a solid mass of mycelium. In some countries it is called a 'Vegetable Caterpillar'.
Habitat and season: Fairly common in woods and meadows and occasionally from the cavities and crannies in old walls. From September to October.
Edibility: No.

* **Cordyceps entomorrhiza** (Fries) Link

A very rare fungus. Originally found one autumn in 1785 on the larva of a beetle at Bulstrode, Buckinghamshire. It does not seem to have been recorded in this country since. The head is 0·5 cm across and at the end of a long, thin and very twisting grey stipe. From 3–8 cm long, may even reach twice that length.

30 Cordyceps capitata (Fries) Link
The Round-headed club

Fruit-body: 3–10 cm high. Tough, thick and yellowish with oval, brown or black head ½ to 1 cm in diameter. The head is roughly textured and may be powdered white by spores.
Spores: 10–50 × 3–6 μm.
Habitat and season: Found mainly in pine woods where it grows on buried truffles just beneath the ground surface. Time September to November, but by no means common. Parasitic on species of *Elaphomyces* (false truffles).
Edibility: Not known.

31 & 32 Cordyceps ophioglossoides (Fries) Link
Adder's tongue club

Fruit-body (clubs): 3–10 cm high. The upper part is brownish black studded with pale points which are the openings of small chambers containing the spores.
Spores: White and smooth, thread-like; part-spores 2·5–5 × 2 μm.
Stipe: Olive to yellowish and cylindrical.
Habitat: Is parasitic on *Elaphomyces muricatus* Fries. The latter may be found in the soil of coniferous plantations and mixed woodlands.
Edibility: No.

Order-*PLECTASCALES*

Illustration 32 shows *C. ophioglossoides* and its host *Elaphomyces muricatus*, in cross-section.
The fruit-body is globose and measures 2–5 cm across. It is hard, bright yellowish-brown with conspicuously pointed warts. In section, it is purplish-brown and marbled. Later the internal mass of the fungus becomes blackish-grey with dust-like gleba.
Habitat and season: *E. muricatus* may be found at any time of the year in soil of pinewoods particularly, and frondose woods occasionally. The best time to search is during September and October when the fruit-bodies of its parasite *Cordyceps ophioglossoides* may be seen above the ground.
Edibility: Of both host and parasite is not known.

OTHER FALSE TRUFFLES

* **Elaphomyces granulatus** Fries

Often parasitised by *Cordyceps capitata* (Fries) Link. Similar in shape and size to *E. muricatus* but finely warted and not marbled in cross section.
Spores: Black and spiny. 24–32 μm.
Habitat and season: Grows in the soil of conifer plantations. Again, the best time to search is September and October.
Edibility: Unknown.

* **Elaphomyces variegatus** Vittadini

Very small, only 0·5 cm in diameter and found in the soil under deciduous trees, especially beech.
Edibility: Unknown.

29 Cordyceps militaris

30 Cordyceps capitata

31 Cordyceps ophioglossoides

32 Cordyceps ophioglossoides and Elaphomyces muricatus

BASIDIOMYCOTINA

APHYLLOPHORALES

Family-THELEPHORACEAE

This is an artificial family which has been considerably reorganised over the last twenty years. It is comprised of species with well-defined hymenium, smooth or wrinkled. The fruit-bodies may be erect and stipitate, cup-shaped or spread out with the upper portion free or altogether resupinate.

They are gelatinous, leathery or waxy in consistency and may be growing on wood or soil.

33 Stereum gausapatum Fries
Shaggy-cloth stereum

Fruit-body: Pileate or resupinate. Forming tiers of small uneven brackets on a vertical surface. 1–4 cm broad, thin, tough and leathery. Upper surface bay-brown or greyish-brown, hispid and lobed. Margin white when young.
Hymenium (lower surface): Smooth and dingy brown to deep chestnut-brown, whitish at the edge. Specimens in prime condition discolour red if cut or bruised.
Spores: White and oblong. 7–10 × 3–3·5 μm.
Flesh: Thin, soft and flexible when young.
Habitat and season: Common all the year round, chiefly on felled or standing oaks. Causes heart rot rendering the timber commercially useless. Not recorded in America.
Edibility: Worthless.

* **Stereum purpureum** (Persoon) Fries
Purple stereum

Similar in form to *S. gausapatum* but upper surface whitish to greyish, woolly or hispid and lobed.
Hymenium: Smooth when young and lilac to purplish. Discolouring with age.
Spores: White and elliptical or oblong. 5–9 × 3–4 μm.
Habitat and season: On fallen or standing trees and branches etc., from January to December. Causes decay in timber and the well-known disease 'silver leaf' in plum trees. Not recorded in America.
Edibility: No.

* **Stereum hirsutum** (Wild) Fries
Hairy stereum

Again similar in form to *S. gausapatum* but upper surface hairy, ochre to greyish and zoned.
Hymenium: Smooth and bright yellow, then fading.
Spores: White and elliptical or oblong. 5–7 × 2·5–3·5 μm.
Habitat and season: parasitic or saprophytic on standing and fallen trees. Occurs all the year round.
Edibility: No.

34 Thelephora terrestris Fries
Thelephora laciniata Persoon ex Fries
Earth fan

Fruit-body: 3–8 cm broad. Fan-shaped and erect but sometimes spread out and irregularly funnel-shaped or circular with upturned margin. Brown to blackish, radially fibrillose and scaly. Margin paler and fringed.
Hymenium: Pale cocoa-coloured, wrinkled or granular.
Spores: Dark brown, warted and angular. 8–9 × 6–7·5 μm.
Flesh: Brown and thin.
Habitat and season: Common, usually in clusters on the ground in pine woods and on heaths. From August to December.
Edibility: No.

* **Thelephora palmata** (Bulliard) Pattouillard

Fruit-body: Larger than *T. terrestris* and less frequently found. From a stem-like base it produces numerous purply-brown, flattened and clustered branches. These are 3–6 cm high with whitish tips, they are leathery and have a foetid odour.
Spores: Reddish-brown in the mass, angular and spiny. 8–11 × 7–8 μm.
Habitat and season: May be found occasionally on the ground in pine woods. August to November.
Edibility: No.

Family-MERULIACEAE

The fruit-body is resupinate or reflexed, soft, waxy and gelatinous at first, but can be membranous or woolly. The hymenium is smooth then reticulated into folds, forming irregular, sometimes toothed pores. The spores are white or coloured. The species are found mainly on wood.

35 Serpula lacrymans (Fries) Karsten
Merulius lacrymans Fries
Dry rot fungus

Fruit-body: Usually resupinate and widely spreading, sometimes forming brackets on a vertical substrate. Spongy flesh, arising from strands of white or greyish mycelium and exuding drops of water when growing.
Pores: Shallow and ferruginous yellow, becoming yellow towards the barren white margin.
Spores: Reddish-rust colour in the mass. Yellow under the microscope. Elliptic 8–10 × 5–6 μm.
Flesh: Thin and greyish-white. Spongy flesh with a fishy smell.
Habitat and season: Mainly on wood in buildings such as roof and floor joists, panelling, skirting boards etc. Particularly in empty structures when there is no air conditioning. It has no respect for timber species and will attack all kinds even the hardest teak, travelling over metal, brickwork and even through mortar in its efforts to

33 Stereum gausapatum

34 Thelephora terrestris

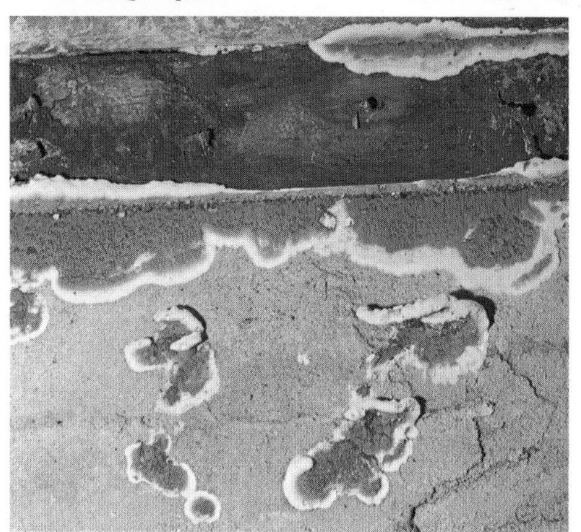
35 Serpula lacrymans

36 Clavariadelphus pistillaris

attack other timber far from the original infection. It will destroy all the inside timber of a building in a comparatively short period of time, and is active in all seasons of the year.
Edibility: No.

* **Merulius tremellosus** Fries

Fruit-body: 1–6 cm across. Resupinate then reflexed. Bracket-like and usually super-imposed, dirty white, woolly, translucent and gelatinous.
Pores: Short, twisted and toothed. Ruddy or pinkish.
Spores: White, smooth and sausage-shaped. 4–5 × 1 μm.
Habitat and season: Fairly common on the underside of fallen branches, stumps etc., often hidden from view. September to March or even later.
Edibility: No.

* **Phlebia merismoides** Fries

Fruit-body: Up to 30 cm across. Thin and resupinate on and following the contour of the substrate on which it grows, with raised vein-like wrinkles or ridges (hymenium) and free margin. Flesh-coloured to purplish with fringed orange coloured margin. Becoming dowdy with age.
Spores: White, smooth, curved and cylindrical, 4–5 × 1·5–2 μm.
Flesh: Tough and gelatinous.
Habitat and season: Common on bark and stumps etc., of frondose trees. Mainly in autumn and winter.
Edibility: No.

Family-CLAVARIACEAE Fairy clubs

Erect fruit-bodies, coralloid, branched or simple. The hymenium more or less covers the whole surface.

The spores are white, yellowish or brown; no cystidia are present.

36 Clavariadelphus pistillaris (Fries) Donk
Clavaria pistillaris Fries

Fruit-body: 8–30 cm high and up to 5 cm thick above. Yellowish, ochraceous or orange; finally dingy brown. Without a distinct stipe. Club-shaped, stuffed, minutely velvety and wrinkled.
Spores: White to slightly ochraceous in the mass. Smooth and pip-shaped. 12–16 × 7–8 μm. Basidia 2- or 4-spored.
Flesh: Whitish, loose and cottony in the centre. Taste bitterish.
Habitat and season: On the ground gregarious or scattered, in frondose or mixed woods. Common in North America but rather rare in Britain. September to December.
Edibility: Of inferior quality.

37 Clavulinopsis pulchra (Peck) Corner
Clavulinopsis laeticolor (Berk. & Curt.) Petersen

Fruit-body: 1·5–10 cm high. Simple and often somewhat flattened, rugulose and glabrous. Fusiform or with a rounded or truncate apex, or branched once with acute tips. Various shades of yellow to bright orange in the fertile area. Always clear white at the very base. Tips discolouring to dirty yellow-ochre with age.
Spores: White, broadly ovate, smooth, slightly thick-walled with a large prominent and long apiculus. 4–7 × 3·5–5 μm.
Flesh: Tasteless and inodorous.
Habitat and season: On deciduous or coniferous leaf debris in autumn. Found in N. America and continent of Europe.
Edibility: No.

38 Clavariadelphus ligula (Fries) Donk
Clavaria ligula [Schaeff.] Fries

Fruit-body: 2–5·5(7) cm high. Simple, cylindrical to clavate, often becoming irregularly flattened with age. Apex subacute to obtuse, becoming rugose. Pale ochraceous-buff, pale salmon or vinaceous-buff. Smooth becoming rugulose, base white tomentose to strigose with abundant white mycelium binding the substrate.
Spores: White in the mass, narrowly ellipsoid, thin-walled and smooth. (10)12–15(18) × (2·5)3–4·5 μm.
Flesh: White and firm, becoming floccose in the apex. Odourless, taste mild to slightly bitter.
Habitat and season: Gregarious to sub-caespitose on coniferous debris. From July to November.
Edibility: No.
Like a small version of *C. pistillaris* but with narrower spores.

39 Clavaria fumosa Fries
Smoky clavaria

Fruit-body: Consists of several worm-like spindles growing from a common base in the form of a close tuft. In many cases the spindles are flattened and have a longitudinal groove. It grows from 4–10 cm high and is greyish yellow at the base shading upwards to a smoky or amethystine yellow.
Spores: Oblong or pip-shaped 6–8 × 3–4 μm. Basidia 4-spored.
Habitat and season: In moist mossy woods and damp mossy lawns. Found occasionally from August to November.
Edibility: Is said to be edible, pleasant taste.

* **Clavaria fusiformis** Fries

Fruit-body: It grows in a similar fashion to the previous species but is of a bright canary yellow with spindles occasionally branched or forked. They also terminate in a point and later become hollow.
Spores: Almost spherical 5–7 μm.
Edibility: Taste bitter.

40 Clavaria argillacea (Persoon) Fries

Fruit-body: 3–8 cm high. Club-shaped and unbranched, blunt or rounded at the apex. Cylindrical or flattened, fragile and often grooved. Colour variable—dirty white, pale greenish-yellow or dirty yellow.
Spores: White, smooth and elliptical. 10–11 × 5–6 μm.
Flesh: Yellowish with a taste resembling tallow.
Habitat and season: Grows loosely tufted in mossy areas on peaty ground. Very common from July to November.
Edibility: Worthless.

37 Clavulinopsis pulchra

38 Clavariadelphus ligula

39 Clavaria fumosa

40 Clavaria argillacea

41 Clavaria fistulosa Fries
Clavariadelphus fistulosus (Fries) Corner

Fruit-body: 10–20 cm high. 0·5 cm thick above. Narrowly club-shaped. Tough in the lower part, fragile at apex. Usually twisted and wavy, becoming hollow. Yellowish, especially below, then brown.
Spores: White, smooth and spindle-shaped. 12–17 × 5·5–7·5 μm.
Flesh: Concolorous with outer surface.
Habitat and season: Uncommon. Grows solitary or in small numbers on wood debris, especially beech. September to February.
Edibility: No.

* **Clavaria contorta** Holmsk
Clavariadelphus contorta

Similar to *C. fistulosa* but much smaller and irregularly shaped. A rare species found on alder branches. Not recorded in N. America.

42 Clavaria juncea Fries
Clavariadelphus juncea (Fries) Corner

Fruit-body: 6–15 cm high and 1 mm thick. Unbranched and thread-like, tapering slightly upwards. Very brittle.
Spores: White and smooth. 7–11·5 × 3·5–5·5 μm.
Habitat and season: A rare species that grows in clusters on damp leaf debris during autumn.
Edibility: No.

43 Clavulina cristata (Fries) Schroeter
Clavaria cristata Fries
Crested clavaria

Fruit-body: 3–8 cm high and irregularly branched. They become flattened upwards and at the apex are cut into acute branchlets like the antlers of a stag. The colour varies from white or pinkish-white to greyish-white. The short, stout base is often blackened by the growth of a parasitic mould *Helminthosphaeria clavariae*.
Spores: White, smooth and elliptical. 7·5–9·5 × 6–7·5 μm pointed at the base and with a large central oil drop. The basidia are 2-spored.
Flesh: White.
Habitat and season: Very common, usually gregarious or crowded, on the ground in woods. June to December.
Edibility: Worthless.
It is very variable and close to *C. cinerea* which is bright grey in colour; transitional forms occur.

* **Clavulina rugosa** (Fries) Schroeter
Clavaria rugosa Fries
Wrinkled clavaria

Fruit-body: 4–12 cm high. Similar to *C. cristata* but is longitudinally wrinkled and has a blunt apex.
Spores: White, smooth and elliptical with a large central oil drop. 9–12 × 7–10 μm. The basidia are 2-spored.

* **Clavaria vermicularis** Fries
White spindles or Worm-like clavaria

Fruit-body: The glistening white clubs of this fungus are cylindrical in shape and pointed at the top. Very fragile in substance and becoming hollow with age. When found in prime condition it is most beautiful and impressive. Height up to 7 cm.
Spores: White, pip-shaped. 4–5 × 3 μm. Basidia 4-spored.
Habitat and season: Can be found growing in clusters in pastures where it likes the shade provided by lush grass, consequently it is not always easy to locate. Found July to November.
Edibility: Yes, but unsubstantial.

44 Ramaria gelatinosa (Coker) Corner var. *oregonensis*
POISONOUS

Fruit-body: 7–12 cm high, 5–14 cm wide. Branched, creamy white, becoming pinkish to deep flesh colour.
Spores: Pale ochre and warty. 7·5–9 × 4·5–5 μm.
Stipe: Not distinct.
Flesh: Yellowish, gelatinous and marbled when cut. Taste acrid then bitter.
Habitat and season: Not a British species. Found in North America under mixed conifers and hardwoods.
Edibility: Considered *poisonous*.

41 Clavaria fistulosa

42 Clavaria juncea

43 Clavulina cristata

44 Ramaria gelatinosa

45 Ramaria aurea (Fries) Quélet
Clavaria aurea Fries
Golden coral fungus

Fruit-body: 7–17 cm high, 5–8 cm wide. Trunk thick, short and elastic. Dividing into straight dichotomous branches. Cream-coloured but yellow upwards.
Branches: Numerous, cylindrical and large, twisted and toothed at the tips. Golden yellow, but later ochraceous yellow.
Spores: Yellowish, 10–12 × 3·5–5·5 μm.
Flesh: Firm, white to yellow.
Habitat and season: In damp situations on the ground in mixed woodlands. From August to November. Occasional.
Edibility: Not recommended as it can be easily mistaken for species which are regarded as poisonous.

46 Ramaria stricta (Fries) Quélet
Upright ramaria

Fruit-body: 3–10 cm high. Trunk thickish, pale brown often with a violet tinge.
Branches: Slender and numerous. Erect, adpressed and parallel. Pale brown, bruising dark. Tips are pale yellow.
Spores: Yellowish and minutely rough. 7–10 × 4–5 μm.
Flesh: Tough and bitter, with a faintly spicy odour.
Habitat and season: Fairly common growing from rotting wood, stumps etc., which are penetrated by the long white mycelial strands. August to January.
Edibility: Worthless.

47 Ramariopsis kunzei (Fries) Donk
Clavaria chionea Persoon

Fruit-body: 3–9 cm high with loosely but repeatedly forked branches of more or less equal length, the branches are cylindrical and curve outwards. It has a short and slender stem-like base which on occasions is pinkish. The whole fungus is very fragile and of shining ivory-white.
Spores: White, smooth and spherical with a small apiculus at the base and a large central oil drop. 3·5–5 μm. Basidia 4-spored.
Flesh: White and brittle.
Habitat and season: This striking and elegant species is locally common, growing in moss or grass, in woods and shady pastures. August to November.
Edibility: Unsubstantial and worthless.

Genus-Sparassis

Fairly large species with fleshy erect fruit-bodies, densely branched and terminating in variously flattened and contorted lobes. The fertile portion is borne on the lower surfaces of the lobes. The spores are white or buff in the mass.

* **Sparassis crispa** Wulfen ex Fries

Fruit-body: Globose 25–35 cm in diameter. At a distance it resembles a large sponge or a cauliflower head. Very densely branched from a very short stout rooting stipe. The branches are much divided and recurved ending at the apices in flattened, wavy, closely packed, crisped lobes. Whitish or pale ochraceous becoming black when old. The hymenium is smooth.
Spores: White, pip-shaped. 5–7 × 4–5 μm.
Stipe: Very short and stout. Whitish but black later.
Flesh: Brittle with a pleasant smell of anise.
Habitat and season: Occasional at the base of pine trees or stumps. From August to November. Often appears in same place year after year.
Edibility: Good when young but must be washed thoroughly.

48 Sparassis radicata Weir

Very similar to *S. crispa*. Found at the base of pines in N. America. Not a British species.
Spores: White, 5–6·5 × 3–3·5 μm.

* **Sparassis laminosa** Fries

Differs from *S. crispa* in its deeper colour, less dense and more irregular habit. Usually found at the base of frondose trees and stumps.

45 Ramaria aurea

46 Ramaria stricta

47 Ramariopsis kunzei

48 Sparassis radicata

Family-CANTHARELLACEAE

The fruit-body is funnel-shaped, usually irregular and lobed. The hymenium is borne on a smooth surface on what could be described as blunt gills or in some cases wrinkles.

The spores are white. They are terrestrial species which grow near to trees. Mainly edible.

49 Cantharellus infundibuliformis (Scop.) Fries
C. tubaeformis Fries

Cap: Dark brown and deeply funnel-shaped with a wavy margin. From 2–5 cm in diameter.
Gills: Vein-like and branched, joined together with cross veins. Yellow at first but later turning grey. Decurrent.
Spores: White, elliptic. 9–11·5 × 6–9 μm.
Stipe: A drab yellow colour 2–8 cm long and often with longitudinal grooves.
Flesh: Tough and bitter.
Habitat and season: Grows on acid soils in mixed woodlands but especially in coniferous woods. Often found in small clusters but also singly. Common from July to December and even January.
Edibility: Of good quality.
The variety (*lutescens*) is entirely yellow with shades of olive in the cap. Also edible.

50 Cantharellus cibarius Fries
The Chanterelle

Cap: 3–10 cm. Top-shaped at first, finally funnel-shaped. The margin is unevenly lobed and crisped. Velvety, then glabrous and dry. The colour is variable but usually some shade of yellow.
Gills (False): Sparse, decurrent and yellow. Thick, blunt and shallow. They are branching and often cross-connected.
Spores: White, elliptical and smooth. 7·5–10·5 × 4·6 μm.
Stipe: Short and stout, narrowing downwards. Concolorous with cap.
Flesh: Tough and fibrous. Yellow, and with the smell of apricot.
Habitat and season: Locally common on the ground in all types of woodland from June to November.
Edibility: Very good and regarded as a delicacy. Young specimens are best and cooking must be long and slow.

Care should be taken not to confuse the above with the species: Luminous clitocybe.

* **Clitocybe olearia** (Fries) Maire POISONOUS
Clitocybe illudens (Schweinitz) Saccardo
Luminous clitocybe

Unlike *C. cibarius*, this is poisonous. It has true gills which are crowded and is found tufted at the base of frondose trees and stumps. Because it is luminous at night it is often called Jack o'Lantern fungus.

51 Cantharellus subalbidus A. H. Smith & Morse
White chanterelle

Cap: 4–12 cm. Lacking a distinct pileus. Top-shaped and stocky, unevenly lobed and crisped, Margin involute. White, becoming rusty-yellow to orange and finally orange-brown where bruised or broken.
Gills (False): Fold-like and close, decurrent. White.
Spores: White and elliptical to broadly elliptical, smooth. 7–9 × 5–5·5 μm. Basidia 4/6-spored.
Stipe: Short and very stout, narrowing slightly downwards. White.
Flesh: Tough and fibrous. White.
Habitat and season: Scattered to gregarious under conifers or in mixed woodland. August to November. Not recorded in Britain.
Edibility: Very good, esteemed in N. America.

52 Craterellus cornucopioides (Fries) Persoon
Cantherellus cornucopioides Fries
Horn of plenty or Trumpet of the dead

Cap: 5–12 cm high. Funnel-shaped, leathery and limp. Dark brown to black when moist, drying out paler. The margin is uneven and crisped.
Hymenium (lower surface): Ash grey, smooth but wrinkled.
Spores: White, elliptical and smooth, 12–15 × 7–9 μm. Basidia 2/4-spored.
Stipe: Short and blackish, merging into the pileus.
Flesh: Tough and blackish-grey.
Habitat and season: Locally common and gregarious. On the ground in frondose woods especially beech. Is often difficult to locate as it is usually all but covered by leaf litter. August to November.
Edibility: Very good grilled with bacon. Alternatively, it can be dried and used for seasoning as it keeps indefinitely.

* **Cantharellus friesii** Quélet

Cap: 2–4 cm. Convex then depressed. Deep orange with at first a yellowish-orange tinge at margin, finally pale opaque buffish-orange.
Gills (false): Decurrent, thick and branched. Very pale pinkish-flesh or with pinkish-lilaceous tinge.
Spores: White, cylindric-ellipsoid, often slightly constricted. 9–12 × 4·5–5(5·5) μm.
Stipe: 16–30 mm long. Solid, silky matt. Continuous with cap. Concolorous or slightly paler than cap with a whitish base.
Flesh: Thin and white, often tinged orange or yellowish in stipe cuticle.
Habitat and season: A rare fungus. On the ground in woods during autumn.
Edibility: Edible but has a sour taste.

49 Cantharellus infundibuliformis

50 Cantharellus cibarius

51 Cantharellus subalbidus

52 Craterellus cornucopioides

53 Polyozellus multiplex (Underwood) Murrill
Cantharellus multiplex Underwood

Cap: 3–5 cm wide. Irregularly infundibuliform with margin lobated, uneven and crisped. Purplish-brown or blackish, the extreme edge having a silvery line, probably from the spores.
Hymenium (lower surface): Greyish-purple with irregular thick and fold-like decurrent gills (false), which are cross-connected by veins and often interrupted by minute slit-like fissures.
Spores: White, rough-walled and copious. Diameter 5–6 μm.
Stipe: 2–4 cm long. Continuous and concolorous with hymenium, blackish base.
Flesh: Tender and brittle. Purple in colour. Taste mild, odour aromatic.
Habitat and season: Densely caespitose and irregular from mutual pressure. On the ground in mixed woods or under conifers. Late summer and autumn. Rare in N. America. Not recorded in Britain.
Edibility: Not known.

54 Cantharellus cinereus (Persoon) Fries

Cap: 3–5 cm high. 1–5 cm across at apex. Irregularly funnel-shaped, thin and flexible with crispate, cracking margin. Brownish-grey in colour.
Hymenium (lower surface): Distinctly grey, with irregular and distant rib-like gills joined by cross veins.
Spores: White and ovoid. 8–10 × 5–7 μm.
Stipe: Brownish-grey. Tapering at base, curved and gradually widening upwards. Grooved and hollow.
Flesh: Black, thin and elastic. Smelling of plums.
Habitat and season: Caespitose on the ground beneath frondose trees, often hidden under fallen leaves. Occasional during August and through to November. In North America it occurs in mixed woodlands.
Edibility: Good but unsubstantial; can be dried.

55 Gomphus clavatus (Fries) S. F. Gray
Neurophyllum clavatum (Fries) Patouillard
Clustered chanterelle or Pig's ear

Fruit-body: 3–16 cm high, club-shaped or turbinate. Flattened or depressed at apex.
Cap: 3–8 cm across, often split down one side, margin thin and lobed. Ochraceous above, violaceous or rosy below and marked longitudinally by veins which develop as the fungus ripens.
Spores: Ochraceous and elliptical. 10–13 × 4–6 μm.
Stipe: Virtually absent.
Flesh: White and aromatic. Some find the taste pleasant, others bitter. Often infected with insect larvae.
Habitat and season: Common in N. America both on the ground and on dead stumps. Caespitose, often many together, under frondose or coniferous trees. From July to November. Rare in Britain.
Edibility: Good but only young specimens should be eaten.

56 Gomphus bonarii (Morse) Singer
Cantharellus bonarii Morse

Fruit-body: Funnel-shaped, up to 25 cm high, in clusters and often branched. Cap surface is set with coarse pyramidal to sub-rectangular scales, often in regular zones, margin involute when young. Orange-yellow, dull ochre, orange-red to scarlet. The scales are usually darker than the ground colour.
Hymenium (lower surface): Consists of decurrent, shallow wrinkled folds, becoming anastomosed and tortuous at apex or not. Creamy or off-white when young, becoming yellowish to near buff with heavy spore deposit later.
Spores: Ochraceous in the mass, ellipsoid to ovoid. 10·5–16(17·5) × 5·5–7(8) μm.
Stipe: Solid when young, becoming internally fibrous with age. Whitish, smooth or minutely hispid, bruising brown.
Flesh: Whitish, except beneath cap surface where it is concolorous with the cap.
Habitat and season: Usually gregarious or clustered under conifers, found in many parts of America and Canada. A variable species with several described forms. Our illustration, the form *bonarii*, is found from central California to British Columbia, Idaho and Montana. Not recorded in Britain.
Edibility: Suspect and not recommended.

53 Polyozellus multiplex

54 Cantharellus cinereus

55 Gomphus clavatus

56 Gomphus bonarii

Family-HYDNACEAE-Tooth Fungi

The species are either stiptate or sessile. The hymenium is inferior in the higher species, superior in resupinate forms, and covered with acute awl-shaped spines or teeth which are distinct at the base.

The spores are hyaline or coloured.

57 Hericium ramosum (Bulliard ex Mérat) Letellier
Hericium laciniatum Leers ex Banker
Bear's head fungus

Fruit-body: Not exhibiting the usual form with a cap and stipe. The sporophores grow from a tough trunk-like attachment on wood, this attachment breaking up into branches which are intricate and fused. Pure white or with a creamy tinge. The whole plant is from 15–30 cm in diameter.
Teeth: Short, up to 5 mm in dense groups hanging from the branches. Pure white, but later tinged with cream.
Spores: White and sub-globose. 3·5–4 × 4–5·5 μm.
Habitat and season: On dead trunks and logs, especially beech. Found during late summer and autumn. Not recorded in Britain.
Edibility: Good.

* **Hericium coralloides** Scop. ex Fries

Similar to *H. ramosum* but with teeth up to 15 mm long which hang from the ends of the branches.
Spores: Smooth and globose. 5–6 μm.
Habitat and season: Found in the autumn in Britain and N. America.
Edibility: Yes.

* **Hericium erinaceus** (Bulliard ex Fries) Persoon

Fruit-body: Solid or porous, soggy, 4 × 8 cm and growing from a tough attachment on wood.
Teeth: Pendulous and neatly crowded, up to 4 cm long. White to yellowish.
Spores: Smooth and globose. 4·5–6 μm.
Habitat and season: Found in the autumn in Britain and N. America.
Edibility: Yes.

58 Hericium abietis (Weir ex Hubert) K. Harrison
Hydnum abietis Hubert E. E.

Fruit-body: Very large, up to 75 cm high and 25 cm wide. Densely branched from a solid thick trunk. The branches are open or compact. White, yellowish or pinkish-buff. Bruising yellow.
Teeth: Up to 1 cm long, pointed. Appearing very short and stout when young. Hanging from the ends of the branches.
Spores: White and sub-globose. 4·5–6 × 4–5 μm. Strongly amyloid.
Habitat and season: Occasional on coniferous logs and trees. Confined to the Pacific North-West of America. Causes a white rot. Not recorded in Britain.
Edibility: Yes.

59 Hydnum imbricatum Fries
Sarcodon imbricatum (Linn. ex Fries) Karsten
Imbricated hydnum

Cap: 5–15 cm sometimes larger. Convex with a central depression, becoming more or less funnel-shaped. The margin is rounded inwardly. Has a greyish background colour and the whole is covered with numerous large, overlapping, concentrically arranged brownish scales which erode from the margin with age.
Spines: Crowded, thin and decurrent. Grey to bluish-grey and up to 12 mm in length.
Spores: Ochraceous, broadly elliptical and minutely warted. 5–7 μm.
Stipe: Short and ample, central or slightly eccentric. Smooth but fibrillose, white to ochraceous-grey.
Flesh: Firm and whitish. Then soft and smoky-grey. Often infected by insect larvae.
Habitat and season: Frequent and gregarious, predominantly in coniferous woods from September to November. Characterised by its scaly pileus and short thick stipe.
Edibility: Good, but only young specimens should be taken. The spines and scales should be removed and the fungus boiled. Pour off the remaining water to remove the bitter taste. If desired it can be preserved in oil or dried after boiling.

* **Hydnum graveolens** Delast

Cap: 3–4 cm across. Flat to depressed undulate. Thin, soft, smooth and leathery. In shades of brown or grey with a pale margin.
Spines: Decurrent, short and greyish-white.
Flesh: Brownish and tough with an extremely strong smell of the clover melilot, which is retained for years when dried.
Habitat and season: Gregarious in pine woods. A very beautiful species when in prime condition.
Edibility: No.

60 Hydnum repandum Fries
Wood-hedgehog or Hedgehog mushroom

Cap: 3–15 cm. Convex, flattened or depressed. Uneven and fragile with incurved margin. Colour whitish to pinkish-buff.
Spines: White or cream, pointed and extending down the stipe.
Spores: Creamy white. 7–8·5 × 6–7 μm.
Stipe: Thick, short and solid. Often eccentric. White or cap colour.

57 Hericium ramosum

58 Hericium abietus

59 Hydnum imbricatum

60 Hydnum repandum

Flesh: White, thick and brittle.
Habitat and season: In groups or rings on ground in woodlands. Common from August to November.
Edibility: Tastes bitter when raw but good after cooking.

* **Hydnum rufescens** Persoon

Similar to *H. repandum* but cap 3–10 cm and pinkish-brown. Spines not decurrent. Grows mainly with conifers.

* **Auriscalpium vulgare** Fries ex S. F. Gray
Hydnum auriscalpium Linnaeus
Ear-pick fungus

Cap: 1–2 cm across. Heart-shaped, thin, leathery and rough. Brown then blackish.
Spines: Tenacious and more or less flesh-coloured.
Spores: White and ovoid. 4–4·5 × 3·5–4 μm.
Stipe: Lateral, slender and hairy. Concolorous with cap.
Flesh: Leathery.
Habitat and season: Fairly common but often overlooked. Found on fallen, decaying pine cones all through the year.
Edibility: No.

61 Hydnellum aurantiacum (Fries) Karsten
Hydnum aurantiacum Bataille ex Fries
Calodon aurantiacum (A & S) Quélet

Cap: 3–15 cm across. Irregularly convex with small protruberances, finely tomentose to velvety, becoming roughened; not zoned. Orange fading out to white at margin, dingy with age.
Spines: Decurrent, whitish changing to pale brown.
Spores: Brown in the mass. 5·5–7·5 × 5–6 μm. Non-amyloid.
Stipe: Variable in length and width. Very firm. Orange becoming bright rusty cinnamon, finally dark brown.
Flesh: Corky and zoned orange to rusty cinnamon. Inodorous.
Habitat and season: Clustered or gregarious under conifers. Late summer and autumn.
Edibility: No.

* **Hydnellum suaveolens** (Fries) Karsten

Cap: 6–15 cm. Irregularly convex to flattish with small protuberances. Pure white and felty at first, then bluish-ochre, finally dingy.
Spines: Crowded, thin, short and decurrent. Bluish then brownish.
Spores: Brown in the mass. 3–3·5 × 3·5–5 μm. Non amyloid.
Stipe: Indigo blue to blue-black. Mycelium deep blue.
Flesh: Corky. Indigo in stipe but paler towards base and in cap. Aniseed odour.
Habitat and season: Solitary or gregarious under conifers during late summer and autumn.
Edibility: No.

* **Hydnellum zonatum** (Fries) Karsten

Cap: 3–6 cm. Depressed or funnel-shaped, thin and leathery. Zoned ferruginous with paler margin.
Spines: Slender, acute and decurrent. Pale then ferruginous.
Spores: Brown, sub-globose, warty and rough. 4·5–6 × 4–4·5 μm.
Stipe: Short and thick. Minutely squamulose, ferruginous.
Flesh: Thin and homogeneous.
Habitat and season: Common in North America, occasional in Britain. In all types of woodland. A variable species.
Edibility: Not known.

62 Hydnellum peckii Banker
Calodon peckii (Banker) Snell & Dick

Cap: 2–15 cm. Club-shaped at first, then turbinate, convex, plane or depressed. Exuding droplets of red juice when young. Tomentose but glabrous with age. White, ageing and bruising reddish-brown.
Teeth: 3–8 mm long, crowded, fine and tough. White, then reddish-brown.
Spores: Brown, sub-globose and tuberculate. 4·5–5·5 × 3·5–4·5 μm.
Stipe: Concolorous with cap or darker. Felty tomentose, attenuating downwards to a hard white mycelial root.
Habitat and season: Gregarious under conifers, especially in wet areas. Often under the same trees year after year. Late summer and autumn. Very variable.
Edibility: No.

* **Hydnellum diabolus** Banker
Hydnellum rhizopes Coker

Differs from *H. peckii* by having a strigose cap surface, matted with age. Not glabrous. Has a strong odour which is sweetish resinous and thinner flesh.
Spores: Sub-globose to irregular. 4·5–6 × 4–5·5 μm.

63 Phellodon tomentosus (Fries) Banker
Hydnum tomentosum Fries

Cap: 2–7 cm. Variously and irregularly shaped. Thin and light brown with strong darker zones, especially away from the margin.
Spines: Short and decurrent. White to greyish-white.
Spores: White, globose and echinulate. 3–4 μm.
Stipe: Slender, central or eccentric, brownish.
Flesh: Thin, tough and fibrous. Odour pleasant.
Habitat and season: Very common, variable and gregarious. Often in fused sheets under conifers. Not recorded in Britain.
Edibility: Worthless.

* **Phellodon melaleucus** (Fries) Karsten
Hydnum melaleucum Fries
Phellodon ellisianus Banker

Cap: 1–6 cm. Variously and irregularly shaped, rigid and dry. Plane to depressed. White tomentose, becoming dark brown or blackish at centre, and silky grey at margin.
Spines: Short, distant when young. Greyish-white to greyish-brown.
Spores: White, sub-globose and echinulate. 3–4·5 μm.
Stipe: Slender and smooth. Dark brown to black, often with a slender brittle root. The stipe may be branched and support several caps.
Flesh: Thin, tough and fibrous.
Habitat and season: Single or gregarious, under pines.
Edibility: Worthless.

Family-POLYPORACEAE-Polypores

A large family with the hymenium lining the surface of the tubes, pore or reticulations. The context may be fleshy, leathery or woody. Either centrally or eccentrically

61 Hydnellum aurantiacum

62 Hydnellum peckii

63 Phellodon tomentosus

64 Inonotus obliquus

stiptate, sessile or resupinate. The majority grow on wood.

The spores are hyaline or coloured, non-amyloid.

64 Inonotus obliquus (Persoon) Pilàt
Tschaga (Russian name)

Fruit-body: Broadly expanded and up to 14 cm. Sessile and resupinate, irregular and patch-like on stumps, dead branches and living trees, especially silver birch. It has a thread-like margin, the adjoining teeth-like pores are round and distinct but towards the centre are greatly lengthened out; lying one upon another in an imbricated manner. The colour is yellow-brown, finally fuscous black.

It is used as a cure for cancer and tumours. Inedible.

65 Ganoderma applanatum (Fries) Karsten
Artist's fungus

Cap: 5–50 cm across. Sessile (without a stipe). Horizontal, semicircular bracket-shaped. Flattish but with humps that are more or less concentric and rising to the point of attachment where it is up to 8 cm thick. Smooth, dull reddish-brown or cocoa-coloured from deposited spores (as illustration). Has a rounded margin which is whitish at first. It becomes very hard with age.
Tubes: Reddish-brown or cinnamon. Distinct from flesh and often stratified.
Pores: Small. Whitish bruising brown. Then cinnamon and finally brown with age.
Spores: Oval and cocoa-brown in the mass. Outer wall smooth and inwardly ornamented. 9–11·5 × 6–8 μm.
Flesh: Chestnut to umber-brown. Fibrous and corky. Thick and very hard.
Habitat and season: On the trunks of felled, fallen or ageing trees, especially beech. (Our illustration shows it on pine, which is unusual.) It causes heart-rot.
Edibility: Worthless.

* **Ganoderma lucidum** (Leysser ex Fries) Karsten

Cap: Shining brown and with a long shiny stipe.

* **Ganoderma laccatum** (Kalch.) Rea

Cap: Covered with yellow resin which reaches to spore surface. Found mainly on beech.

66 Fomes fomentarius (Fries) Kickx
Tinder fungus

Cap: 6–50 cm. Sessile and broadly attached. More or less hoof-shaped, not flattened but concentrically furrowed with a blunt margin. Pale brown to greyish-brown, becoming grey or greyish-black. Velvety then glabrous.
Tubes: Long and rusty. Multi-layered in old specimens.
Pores: Small and round. Cinereous then ochraceous.
Spores: White. 15–18 × 5·5–6 μm.
Flesh: Corky and tough. Reddish-brown.
Habitat and season: On living or dead latifoliate trees, especially birch, causing white heart-rot. Scotland, Canada and U.S.A. Common all the year round. Individual fruit-bodies live and increase in size over several years.
Edibility: No, but may be used for decorative purposes in the household, when dry.

* **Phellinus igniarius** (Fries) Quélet
Willow fomes

A rust-brown, woody fungus that grows on willows causing heart-rot. The tubes and flesh are brown. It has a white spore-print. The spores are spherical and hyaline. 5–6 μm.

67 Polyporus tomentosus Fries
Coltricia tomentosa (Fries) Murrill
Polyporus aduncus Lloyd
Polyporus peakensis Lloyd

Cap: 3–18 cm broad and 0·5–4 cm thick. Circular or fan-shaped. Whitish when young becoming ochraceous-tan or rust-brown. Extreme margin whitish. Tomentose.
Tubes: 1·5–7 mm long.
Pores: Hoary or brown, bruising darker. Mouths angular to somewhat irregular. Dentate and slightly pubescent.
Spores: Hyaline or very pale coloured under the microscope. Smooth, ellipsoid or oblong-ellipsoid. 4–6 × 3–5 μm.
Stipe: When present, lateral, eccentric or central. Up to 5 cm long or sessile, according to habitat. Ochraceous to dark brown, tomentose like the cap.
Flesh: Spongy and watery when fresh, drying rigid. Concolorous with cap.
Habitat and season: Frequent on stumps, trunks or roots of living or dead conifers. Also on the ground attached to buried wood. Not recorded in Britain.
Edibility: Worthless.

68 Coltricia perennis (Linn. ex Fries) Murrill
Polyporus perennis Fries
Polystictus perennis (Linn. ex Fries) Karsten
Perennial polypore

Cap: 3–9 cm. Funnel-shaped or expanded and only umbilicate. Thin, tough, velvety. Yellowish, rusty or brown. Margin wavy. It is marked with small raised radiating lines giving it a striated appearance. Becoming smooth and zoned.
Tubes: Short, decurrent and cinnamon-coloured.
Pores: Minute, angular or rounded. At first covered with a whitish bloom, but finally becoming dark brown.
Spores: Yellowish to yellow-brown. Elliptical usually with an oil-drop. 6–9 × 4–6 μm.
Stipe: 2–4 cm long. Attenuated below. Downy and concolorous with cap.
Flesh: Thin, fibrous, tough and tawny brown.
Habitat and season: On the ground in woods and heaths, especially where timber has been burnt. Solitary or a few together. May survive for several years. Found all the year round.
Edibility: Worthless.

65 Ganoderma applanatum

66 Fomes fomentarius

67 Polyporus tomentosus

68 Coltricia perennis

69 Piptoporus betulinus (Fries) Karsten
Polyporus betulinus Fries
Birch bracket

Cap: 5–20 cm. Hoof or shell-shaped, extending horizontally from the trunks or branches of dead or dying trees. The smooth, greyish or brownish upper surface is convex with the margin incurved. The smooth surface is formed by a thin cuticle which peels easily in young specimens but later cracks and breaks up leaving the surface mottled and streaked with white. The underside is flat and white, becoming dingy with age.
Pores: White, round, short and minute. Later seceding.
Spores: White 4·5–5·5 × 1·5–2 μm.
Stipe: Short and eccentric or absent.
Flesh: White, thick and elastic, then corky in texture; not woody. Was once used for razor strops. Also ideal for mounting entomological specimens.
Habitat and season: Always confined to dead or dying birch trees. It causes a red rot in the sapwood. A very common species: the fruit-body can be found at all times of the year but new growths usually appear in the autumn.
Edibility: No.

70 Coriolus versicolor (Fries) Quélet
Trametes versicolor (Linn. ex Fries) Pilát
Polystictus versicolor (Linn. ex Fries) Fries
Many zoned polypore

Brackets: 3–5 cm across. Semicircular, flattened, thin and tough. Flexible when young. Usually in tiers and spread end to end along branches. The upper surface is velvety and attractively marked with concentric zones of varying colours—brown, yellow, grey, greenish or black. The margin is usually wavy.
Tubes: White and very short.
Pores: White to yellowish and smooth. Round or angular.
Spores: White, oblong and cylindric. 4–6 × 2–2·5 μm.
Stipe: Absent.
Flesh: White, thin and tough.
Habitat and season: On logs, stumps and branches of frondose trees. A very common fungus which can be found all the year round. It causes decay in stored and felled timber.
Edibility: No.

If the tubes and pores are found to be bright orange or vermillion, the cause of this is likely to be the growth of *Hypomyces aurantius*, a parasitic ascomycete.

71 Meripilus giganteus (Fries) Karsten
Polyporus giganteus Fries
Grifola gigantea (Persoon ex Fries) Pilát
The Great polypore

Cap: Each 10–40 cm wide and in the form of numerous fan-shaped, overlapping brackets. Dark brown with darker lines and zones of hair-like scales.
Pores: Minute and more or less round. Whitish bruising black.
Spores: White and elliptic. 5–6 × 3·5–5·5 μm.
Stipes: Whitish and all uniting into a common, short tuber.
Flesh: Ample and brittle, then leathery. Whitish bruising black. Sour smelling.
Habitat and season: Common at the base of living or dead frondose trees. A handsome species found from July to January.
Edibility: Worthless.

72 Lenzites betulina Fries
Trametes betulina (Fries) Pilát

Bracket: 4–8 cm across. Sessile, semicircular, sometimes almost circular. Flattened and tomentose. Grey or greyish brown, zoned. Margin sometimes wavy. Often green with unicellular algae.
Hymenium (lower surface): In the form of elongated gill-like plates radiating from the expanded base attachment, simple or branched, often anastomising. Greyish-white, with or without yellowish tinge.
Spores: White and cylindrical, 4–6 × 2–2·5 μm.
Flesh: White and thin, soft to corky.
Habitat and season: Common on stumps and branches of frondose trees especially birch. Often tiered and sometimes resupinate. January to December.
Edibility: No.

* **Lenzites saepiaria** Fries
Gloeophyllum sepiarium (Wulfen ex Fries) Karsten

Bracket: 3–6 cm across. Horizontal, hard, strigose or tomentose, wrinkled, finally squamulose. Bay or umber and zoned. Margin yellowish.
Hymenium (lower surface): Gill-like, branched and anastomosing. Very rigid. Yellowish then umber.
Spores: White, cylindrical, smooth and slightly curved. 8–11 × 3–4 μm.
Flesh: Thick and tough. Brownish-yellow.
Habitat and season: Occasional on conifer trees and stumps, also on worked timber in dwellings. Causes serious decay.
Edibility: No.

69 Piptoporus betulinus

70 Coriolus versicolor

71 Meripilus giganteus

72 Lenzites betulina

73 Laetiporus sulphureus (Fries) Murrill
Grifola sulphurea Bulliard ex Fries
Polyporus sulphureus Bulliard ex Fries
Chicken of the woods or Sulphur polypore

Cap: 30–40 cm. Fan or tongue-shaped with the margin undulate and lobed. The whole extending horizontally from the substrate. Cap colour at first orange-yellow or rosy, with a sulphur yellow margin. Later fading, becoming dull and dirty.
Tubes: Short and sulphur yellow.
Pores: Small and round. Sulphur yellow and exuding dewy droplets when young.
Spores: White and elliptical. 5–7 × 4–5 μm.
Stipe: Very short or absent as such. The many overlapping brackets arising from a common base on the host.
Flesh: Yellow and then dirty white. Acrid, soft and cheesy when young, exuding yellow milk when broken. Later becomes dirty white, dry and brittle.
Habitat and season: Common on living or dead trees from June to October. It seems to show a preference for oak. A wound parasite causing a reddish-brown heart-rot to which the host gradually succumbs. Like *Polyporus squamosus*, it produces staghorn-like fructifications when growing in dark situations.
Edibility: Can be eaten when young but certainly not recommended.

74 Polyporus squamosus Fries
Dryad's saddle or Scaly polypore

Cap: 10–30 cm. Semi-circular or fan-shaped. Horizontal and flattened. Ochre yellow above and variegated with broad adpressed dark brown feathery scales, more or less concentrically arranged.
Tubes: White to yellowish and decurrent. Up to 7 mm long.
Pores: Minute, then large angular and torn. Whitish.
Spores: White, long and elliptical 10–15 × 4–5 μm.
Flesh: White. Soft at first then tough.
Habitat and season: Common, usually gregarious and tiered on trunks of living or dead trees, especially elm and sycamore. It is a wound parasite and causes heart-rot. When growing in dark places, no pileus is produced but the stipe branches and elongates into staghorn-like structures.
Edibility: Not recommended but non-poisonous.

75 Polyporus brumalis Fries
Winter polypore

Cap: 2–5 cm in diameter, convex at first but finally depressed. Greyish to light brown in colour. Older specimens sometimes develop darker concentric zones.
Tubes: Dirty white, short and decurrent on stipe.
Spores: White to yellowish, oblong, 5–8 × 1·5–2·5 μm.
Stipe: Greyish to brown tending to be darker at the base. Central or slightly eccentric and very tough.
Habitat and season: Attached to wood of frondose trees, stumps and roots etc. When on buried wood appears to be growing from the ground and resembles a small Boletus. Grows singly or a few together. Appears at all times of the year and is fairly common.
Edibility: Worthless, the flesh being extremely tough.

* **Polyporus varius** Fries

Cap: 2–10 cm across. Circular to shell-shaped, smooth and tough. Often depressed at the point of origin of the stipe. Brownish-yellow, usually with fine radiating lines. Soon becoming woody and rigid. A variable species.
Tubes: Shallow and decurrent. White to creamy.
Pores: More or less round, minute. Cream colour becoming grey with age.
Spores: White and oblong. 6·5–8 × 3–3·5 μm.
Stipe: Eccentric or lateral, smooth and usually short. White or yellowish, becoming darker at base.
Flesh: Thin and very tough. White to dirty yellowish.
Habitat and season: Common trunks, stumps and fallen branches of frondose trees, especially ash. July to November.
Edibility: Worthless.

* **Polyporus nummularius** Fries

Cap: Paler and smaller than *P. varius* 2–4 cm across.
Spores: White. 7–9 × 2·5–3 μm.
Stipe: Slender and blackish.
Habitat and season: Often growing on *Salix* (Willows). July to November.
Edibility: Worthless.

76 Fistulina hepatica Fries
Beef steak fungus

Bracket: 5–30 cm shaped like a hoof or tongue. Reddish-brown and sticky on top.
Tubes: Minute and yellowish flushed reddish. Pore mouths are pale flesh coloured and round.
Spores: Pinkish-brown and egg shaped. 4–6 × 3–4 μm.
Stipe: Virtually absent.
Flesh: Red, fibrous and veined (resembles red meat). It exudes a reddish juice when cut.
Habitat and season: Fairly common on the trunks of living frondose trees, especially oak. It causes the wood to become rich brown in colour before decay. August to November.
Edibility: Can be eaten sliced and cooked in the same way as beef steaks or can be used raw in salads. The flavour is acidulous and its reputation much overrated.

73 **Laetiporus sulphureus**

74 **Polyporus squamosus**

75 **Polyporus brumalis**

76 **Fistulina hepatica**

Order-*AGARICALES-GILL FUNGI AND BOLETI*

Family-AGARICACEAE-Gill Fungi

Genus-*Hygrophorus* (moisture bearing)

Medium to small species, all of which are terrestrial in grassland or woodland. The caps, especially in young specimens, are often brightly coloured and viscid, with watery flesh, decaying and discolouring quickly. The gills are waxy, distant and thick, but acute at the edge. Adnate or adnato-decurrent, often branched and veined. The spores are white, smooth and non amyloid. The stipe is fleshy and central, the flesh being continuous with that of the cap.

For the most part, the species in this genus are edible, some are considered good, a very few may cause indigestion in allergic people.

The *Hygrophori* are often divided into three subgenera:

Hygrocybe: Veil absent, fruit-body slender, watery and fragile. Cap mainly brightly coloured, thin and fragile. Slimy in moist conditions, shining when dry. The stipe is hollow, smooth or fibrillose, not ornamented. Grassland.

Camarophyllus: Veil absent. Cap firm and opaque, never viscid. The stipe is even, fibrillose or glabrous. Not ornamented at apex with dots. Includes white or sombre coloured species. Grass or parkland.

Hygrophorus (=Limacium): Cap viscid. Universal veil viscid. Partial veil floccose, sometimes forming trace of a ring, or attached to cap margin. Stipe ornamented with scales or squamules near apex. Woodland.

Subgenus-*Hygrophorus*

77 Hygrophorus olivaceo-albus (Fries) Fries
Limacium olivaceo-album (Fries) Kummer

Cap: 3–8 cm. Olive-brown becoming paler especially towards margin. Fleshy at disc but otherwise thin. Conical then plane. Covered with olive gluten which disappears with age; the margin then appears striate.
Gills: Slightly decurrent and distant. Shining white or stained olivaceous by the gluten. Connected by veins at the base.
Spores: White and elliptical. 7–8 × 4 μm.
Stipe: Long, slender, solid and viscid. Whitish and sheathed below with spotted squamules from the brown veil which terminates in a ring-like zone near to apex. Above this the stipe is shining white.
Flesh: White.
Habitat and season: Fairly common on the ground in woods and woodland pastures, mainly coniferous. Common in N. America but rare in Britain.
Edibility: Yes, good.

* **Hygrophorus hypothejus** (Fries) Fries
Limacium hypothejum (Fries) Kummer
Herald of the Winter

Similar to *H. olivaceo-albus* but has yellow gills and is usually a late species.
Spores: Ovate 7–9 × 4–5 μm.

* **Hygrophorus agathosmus** (Secretan) Fries
Limacium agathosmum (Secretan) Wünsche

Has a viscid grey coloured cap and is a species of pine woods. Gills and stipe are whitish. Smells strongly of anise.
Spores: Ovoid. 9–11 × 5–7 μm.

78 Hygrophorus chrysaspis Métrod
Limacium melizeum (Fries) Ricken

Cap: 4–8 cm. Convex to umbonate, and glutinous. White, yellowing with age.
Gills: White to pale orange. Distant and decurrent, sometimes veined near the stipe.
Spores: White or slightly tinted pinkish, oval or oblong. 7·5–9 × 4–5·5 μm.
Stipe: Slender and usually tapering at base. Viscid. Granular at apex.
Flesh: White, firm and brittle, thick at umbo. Has an odour of formic acid. Strong reaction with Potassium hydroxide KOH.
Habitat and season: Common and gregarious in frondose woods, especially beech. From August to November.
Edibility: Yes, but of poor quality.

* **Hygrophorus eburneus** (Fries) Fries

Very similar to *H. chrysaspis* but no reaction to KOH.

* **Hygrophorus cossus** (Berkeley) Fries

Again similar to *H. chrysaspis* but evil smelling, like the Goat Moth larva.

Subgenus-*Hygrocybe*

79 Hygrophorus chlorophanus (Fries) Fries
Hygrocybe chlorophana (Fries) Wünsche
Yellow wax-cap

Cap: 2–5 cm. Convex and then flat, often cracking. Margin may or may not be striate. Viscid. Lemon-yellow, paling at centre with age.
Gills: Distant and thin. Adnexed and ventricose. Pale whitish-yellow, then sulphur yellow.
Spores: White and broadly elliptical. 7–9 × 4–5 μm.
Stipe: Smooth and hollow, often compressed. Viscid when moist, shining when dry. Concolorous with cap.
Flesh: Yellow and watery. The whole plant is some shade of lemon-yellow and does not bruise black.
Habitat and season: Common in grassy places. From

77 Hygrophorus olivaceo-albus

78 Hygrophorus chrysaspis

79 Hygrophorus chlorophanus

80 Hygrophorus miniatus

August to November.
Edibility: Not known.

⋆ **Hygrophorus obrusseus** (Fries) Fries
Hygrocybe obrussea (Fries) Wünsche

Similar to *H. chlorophanus* but of stouter habit and cap never striate.
Gills: Broad, thick and becoming free.

⋆ **Hygrophorus flavescens** (Kauffman) A. H. Smith & Hesler
Hygrocybe flavescens (Kauffman) Singer

Also similar to *H. chlorophanus* but has a slimy golden-yellow cap and stipe.

80 Hygrophorus miniatus (Fries) Fries
Hygrophorus miniata (Fries) Kummer
Red-lead hygrophorus

Cap: 1–2 cm. Convex then flat, sometimes umbilicate. Smooth and dry or somewhat squamulose. Crimson becoming paler.
Gills: Adnate, not decurrent. Distant and broad. Orange to crimson.
Spores: White and oval. 7·5 × 5 μm.
Stipe: Slender and even; smooth and dry; shining and crimson.
Flesh: Reddish but paling.
Habitat and season: Gregarious in grass and moss, especially on peaty ground. From July to October.
Edibility: Worthless.

* **Hygrophorus reai** Maire

Similar to *H, miniatus* but is viscid and has a bitter taste. Flesh is orange. Basidia 2- or 4-spored.
Spores: Elliptical. 7–8–10 × 4·5–4·75 μm.

81 Hygrophorus calyptraeformis Berkeley & Broome
Hygrocybe calyptraeformis (Berkeley & Broome) Fayod
Ballerina hygrophorus

Cap: 4–7 cm high and the same across when expanded. At first acutely conical then expanded with centre remaining acutely umbonate. Edge becomes upturned with extensive splitting. Fragile. Rose-pink in colour becoming paler with age.
Gills: Distant and narrowed behind. Adnexed or free. Rose-pink and then whitish.
Spores: White and elliptical. 7–8 × 4·5–5 μm.
Stipe: Thick, brittle and hollow. Striate and rather long, splitting lengthways.
Flesh: Pink in the cap and white in the stipe. Does not bruise black.
Habitat and season: Fairly common in pastures, lawns and heaths. Found from late summer to autumn. A beautiful species which is readily identified by its colour, form and fragility. Not recorded in N. America.
Edibility: Yes.

82 Hygrophorus coccineus (Fries) Fries
Hygrocybe coccinea (Fries) Kummer
Scarlet wax-cap

Cap: 2–5 cm across. Bell-shaped at first but becoming flat. Vivid shiny red when young but later fading to yellow.
Gills: Orange-yellow flushed with red. Adnate and decurrent by a tooth.
Spores: White, oblong or elliptic. 7–9·5 × 4·5–5·5 μm.
Stipe: Hollow. Yellow below gills but shading to red and then back to yellow again towards the base.
Flesh: Red or yellowish.
Habitat and season: Mossy and grassy places such as lawns and pastures. Common from June to December.
Edibility: Worthless.

83 Hygrophorus psittacinus (Fries) Fries
Hygrocybe psittacina (Fries) Wünsche
Parrot hygrophorus or Parrot wax-cap

Cap: 1–1½ cm across. Slimy and varying in colour. Deep turquoise when fresh. Yellow tints appearing with age and often becoming completely yellow ochre. Bell-shaped at first but later expanding with a central umbo.
Gills: Adnate, broad and distant. Pale green shading to yellow at gill edge.
Spores: White, ovate. 8–9·5 × 4–5·5 μm.
Stipe: Slimy and varying in colour just as the cap. Always remains green at apex.
Flesh: In cap whitish tinged with green. Yellowish in stipe.
Habitat and season: Common in grassy places, meadows, hill pastures and lawns. From July to November.
Edibility: No.

84 Hygrophorus conicus (Fries) Fries
Hygrocybe conica (Fries) Kummer
Conical wax-cap or Conical hygrophorus

Cap: 1–5 cm. Acutely conical, finally expanded with a split and uneven margin. Viscid. Yellowish-orange, orangey-red or scarlet; eventually blackening.
Gills: Whitish-yellow, broad, distant and free. They exude a water-like liquid if cut.
Spores: White and ovate. 11–13 × 5·5–6·5 μm.
Stipe: Hollow, cylindrical, brittle and dry. Yellow with a reddish flush. Blackens with age or when bruised.
Flesh: Yellow in cap and whitish in stipe. Bruising black.
Habitat and season: Very common in pastures, lawns and heaths. From July to November.
Edibility: No.

* **Hygrophorus nigrescens** (Quélet) Quélet
Hygrocybe pseudoconica J. Lange

Very similar in appearance to *H. conicus* but the base of the stipe is white.
Spores: Ovoid. 8–11 × 5–6 μm.

* **Hygrophorus laetus** (Fries) Fries
Hygrocybe laeta (Fries) Kummer
Hygrophorus houghtonii Berk. & Br.

Cap: 1–3 cm. Convex then plane, very glutinous. Margin pellucid-striate. Flesh colour to brownish at first, fading to yellow ochre with age.
Gills: Distant, decurrent and connected by veins. Greyish to flesh colour.
Spores: White and elliptical. 5–7 × 4 μm.
Stipe: Long, slender, equal and glutinous. Concolorous with cap except at apex which is bluish-grey.
Flesh: Thin and slightly paler than cap in colour.
Habitat and season: Fairly common in upland pastures and heaths, especially in damp areas. September to November. The slender form, thin cap and bluish-grey stipe apex are characteristic.
Edibility: Good.

81 Hygrophorus calyptraeformis

82 Hygrophorus coccineus

83 Hygrophorus psittacinus

84 Hygrophorus conicus

85 Hygrophorus glutinipes (J. Lange) P. D. Orton
Hygrocybe citrina var. *glutinipes* J. Lange

Cap: 1–2·5 cm. Convex but later depressed. Margin striate. Viscid and entirely yellow.
Gills: Thick and distant. Whitish-yellow.
Spores: White and smooth. 7–9 × 4–5 μm.
Stipe: Yellow and viscid.
Habitat and season: Damp places in pastures and copses, found growing singly or in groups. September to November. Not recorded in America.
Edibility: Not known.
Our illustration shows a specimen in prime condition, the colour quickly fades.

Subgenus-Camarophyllus

86 Hygrophorus pratensis (Fries) Fries
Hygrocybe pratensis (Fries) Donk
Buff cap or Butter mushroom.

Cap: 2–7 cm. Compact and somewhat top-shaped, slightly umbonate. Smooth and dry, margin thin. Colour variable, from pale yellowish to buff or even reddish. Fades with age.
Gills: Thick, broad and distant. Cross-connected and decurrent. They are white at first but later showing tint of cap colour.
Spores: White, smooth and elliptic. 6–8 × 4–5·5 μm.
Stipe: Stout, slightly striate and often curving, with a pointed base. White at first, then pale cap colour. Stuffed but finally hollow.
Flesh: White to pale buff, thick and firm.
Habitat and season: A common species in pastures often growing gregariously. Can be found from August to December.
Edibility: Good to eat but should be cooked slowly.

* **Hygrophorus niveus** (Scop.) Fries
Camarophyllus niveus (Scop.) Fries
Snow-white hygrophorus

Cap: 1–5 cm. Campanulate then convex to expanded. Never funnel-shaped, not fleshy at disc. Margin striate when moist. Pure white.
Gills: Distant, decurrent, thin and broad, sometimes connected by veins. White.
Spores: White, elliptical or pip-shaped. 8–10 × 4·5–6 μm.
Stipe: Slender and equal or slightly attenuated towards the base. Concolorous with cap.
Flesh: Thin and everywhere equal. White and odourless.
Habitat and season: Common in pastures, meadows and heaths. September to November.
Edibility: Good.

* **Hygrophorus russocoriaceus** Berkeley and Miller

This is very similar to *H. niveus* but the gills are usually thicker and with a strong pleasant odour variously described as of Russian leather, Potentilla, Indian Sandal Wood and incense.
Spores: White and elliptical. 7–9 × 4·5 μm.
Edibility: Probable.

* **Hygrophorus virgineus** (Fries) Fries

Again similar to *H. niveus* but larger.
Cap: 4–7·5 cm. White.
Gills: Thick and white.
Spores: White, oblong and elliptical. 9–12 × 5–6 μm.
Edibility: Probable.

* **Hygrophorus lacmus** ([Schum.] Fries) Kalchbr
Hygrophorus subradiatus variety *lacmus* (Schum.) Fries

Cap: 2·5–5 cm. Convex to umbonate and quite fleshy, often squamulose at disc. Striate when moist, shining when dry. Grey or lilac, pallid with age.
Gills: Distant, thin, decurrent and connected by veins. Grey in colour.
Spores: White and nearly spherical. 6·5–7·5 × 5·5–6 μm.
Stipe: Attenuated both at base and apex, often twisted. White to grey but with a definite yellow base.
Flesh: Grey in cap and upper stipe, yellow in stipe base.
Habitat and season: Fairly common in heaths and pasture. From September to October.
Edibility: Good.

87 Hygrophorus purpurascens (Fries) Fries
Limacium purpurascens (Fries) Kummer
Agaricus purpurascens Fries

Cap: 6–12(15) cm. Convex then plane, fibrillose. Viscid when young, margin involute until late maturity. Ground colour whitish, often streaked purplish-red. Fibrils red or deep livid brown at the disc and pinkish at the margin.
Gills: Adnate to decurrent, narrow, crowded to subdistant. White, then pink to purplish-red.
Spores: White, smooth and elliptical. 5·5–8 × 3–4·5 μm. Basidia 2- or 4-spored.
Stipe: Solid, attenuated below, dry. Silky at the apex, more or less concolorous with cap below and often spotted with dark purplish-red.
Ring: Fibrillose and fugacious. White becoming purplish-red.
Flesh: White and firm. Thick at disc, thin at margin. Odour and taste mild.
Habitat and season: Fairly common under conifers from May to December in N. America and continent of Europe. Not recorded in Britain.
Edibility: Not known.

88 Hygrophorus speciosus Peck

Cap: 2–5 cm. Umbonato-convex with margin incurved at first. Smooth and viscid. Orange-red to orange at the

85 Hygrophorus glutinipes

86 Hygrophorus pratensis

87 Hygrophorus purpurascens

88 Hygrophorus speciosus

centre, paler at the margin.
Gills: Moderately distant, adnato-decurrent. White to yellowish.
Spores: White, elliptical and smooth. 8–10 × 4·5–6 μm.
Stipe: Ample, slightly thicker at base. Viscid. White with dull orange stains.

Flesh: White to yellowish and fairly thick.
Habitat and season: Not recorded in Britain. Common in North America, where it grows gregariously or in clusters under conifers, especially larch. From September to November.
Edibility: Not known.

Genus-*Asterophora* (Nyctalis)

Small species. The cap is covered with powdery chlamydospores. The gills are distant, very thick or absent. The stipe is central. It is found growing on decaying species of Lactarius or Russula, especially *R. adusta* and *R. nigricans*.

89 Asterophora parasitica (Fries) Singer
Nyctalis parasitica (Fries) Fries
Pick-a-back toadstool

In the illustration we show it growing on decaying *Russula adusta* (Fries) Fries
Cap: 1–2 cm. Convex, then expanded, pruinose, whitish then grey.
Gills: Thick, distant, adnate, white and fuscous. Finally contorted and anastomising, being obscured by brownish powder-like chlamydospores.
Chlamydospores: Produced on the gills. Pale brown, oval and smooth. 15 × 10 μm. Thick-walled.
Stipe: 2–4 cm long and often curved. White to grey.
Flesh: Brown and thin, with a nauseous odour.
Habitat and season: Clustered on various species of decaying Lactarius and Russula. Found in autumn.
Edibility: No.

* **Asterophora lycoperdoides** (Mérat) S. F. Gray
Nyctalis asterophora Fries

Cap: 1–2 cm. Hemispherical and somewhat conical. White then covered with mealy fawn-coloured coating of chlamydospores.
Gills: Whitish, obscure and shallow; often wanting.
Chlamydospores: Fawn colour in the mass, large and bluntly stellate. 12–15 μm.
Stipe: Very short and whitish.
Flesh: Thick and whitish. Odour strongly rancid.
Habitat and season: On various species of decaying Lactarius and Russula. During autumn.
Edibility: No.

Genus-*Lyophyllum*

Formerly referred to under Clitocybe, Tricholma or Collybia.

All species have basidia with numerous granules, staining deeply when heated with aceto-carmine.

90 Lyophyllum palustre (Peck) Singer
Tephrocybe palustris (Peck) Donk
Collybia palustris (Peck) A. H. Smith
Collybia leucomyosotis (Cooke & W. G. Smith) Saccardo

Cap: 1–2 cm. Convex then expanded, finally depressed. Hygrophanous and striate. Yellowish or greyish-brown with paler margin in young specimens.
Gills: Distant and adnate with a tooth. Dirty white.
Spores: White, ovate and smooth. 6–8·5 × 4–5 μm.
Stipe: Long, slender and tough, hairy at base. Has small whitish scales when young. Colour as cap.
Flesh: Thin and yellowish-brown.
Habitat and season: Common in marshy land where it grows attached to Sphagnum. May to October.
Edibility: Worthless.

91 & 92 Lyophyllum decastes (Fries) Singer
Clitocybe decastes (Fries) Kummer
Tricholoma aggregatum (Secretan) Constantin & Dufour
Fried chicken mushroom

Cap: 4–12 cm. Umbonato-convex with involute margin, then plane and often undulate. Colour variable, grey through to brown or yellowish-brown (see illustrations).
Gills: Adnato-decurrent and rather crowded. White at first then grey to straw-coloured.
Spores: White and almost round. 5·5 × 5–6 μm.
Stipe: Long or short, often curved and slightly eccentric. White or greyish-white.
Flesh: Thin in cap, fibrous-elastic. Whitish. Odour agreeable and mild tasting.
Habitat and season: Common in woods and parkland where it can be found growing densely tufted around stumps; or in grass attached to buried roots. From July to October.
Edibility: Fairly good.

* **Lyophyllum semitale** (Fries) Kühner
Collybia semitalis (Fries) Quélet

Cap: 2·5–10 cm. Convex then plane, glabrous, moist until mature and hygrophanous. Margin incurved when young and pellucidly striate. Dark greyish-brown to dark brown when fully moistened, drying out to greyish-yellow or pale ochre.
Gills: Sub-distant, broad and blunted behind, adnate with a decurrent tooth. Dirty white then darker with age.
Spores: White and oblong. 6·5–9·5 × 4–5 μm.
Stipe: Either long or short, tough and elastic, fibrillosely striate. Base slightly bulbous and often with a root. Brown or greyish-brown.
Flesh: Cartilaginous and often cracking. Grey brown.
Habitat and season: Uncommon, singly or sub-caespitose on the ground in coniferous woods or by paths, especially in very wet weather.
Edibility: Suspicious.

89 Asterophora parasitica

90 Lyophyllum palustre

91 Lyophyllum decastes

92 Lyophyllum decastes

Genus-*Tricholoma*

A very large genus. The cap is fleshy, viscid or dry. The stipe is central, usually fleshy and fibrous, lacking volva, and only rarely displaying ring or veil. The gill attachment varies but usually sinuate. The spores are white or pale pink, smooth with central oil drop, typically non-amyloid.

With very few exceptions they are terrestrial species found chiefly under foliage but a few do occur in open areas such as pastures and lawns etc. A few species can be rated as excellent to eat, many are good, a very few are 'suspect'.

93 Tricholoma caligatum (Viviani) Ricken
Armillaria caligata (Viviani) Gillet

Cap: 8–18 cm. Hemispheric at first and silky squamulose. Margin white, involute and connected to the stipe by the white partial veil. Later expanded and margin often festooned with velar remains. Finally plane, sometimes irregular and cracking when the white flesh becomes visible. Variable colours: yellowish, ochraceous, reddish-brown or totally dark brown with age.
Gills: Crowded, narrow, sinuate and white. Eroding and becoming free.
Spores: White, elliptical and smooth. 6–7·5 × 4·5–5·5 μm.
Stipe: Tall and ample. White and mealy above the ring, brown-squamulose below.
Ring: White and membranous.
Flesh: Thick, firm and white. Odour and taste pleasant.
Habitat and season: Occasional or gregarious under pines. September to November.
Edibility: Excellent.

* **Tricholoma matsutake** Ito & Imai
Armillaria matsutake Ito & Imai

A species found in Japan.

* **Tricholoma ponderosum** (Peck) Saccardo
Armillaria ponderosa (Peck) Saccardo

A North American species.

Both *T. matsutake* and *T. ponderosum* are very similar to *T. caligatum* and considered by some authorities as the same species.

94 Tricholoma saponaceum variety *squamosum* (Cooke) Rea
Soap tricholoma

Cap: 4–8 cm. Convex to umbonate. Then plane, irregular, dry and often cracked. Dull greyish-brown or olivaceous-brown, darker at disc. Becoming squamulose when mature.
Gills: Distant and sinuate. White or with a glaucous tinge, bruising or spotting reddish.
Spores: White and broadly elliptical. 5–6 × 3·5–4 μm. Non-amyloid.
Stipe: Ample, whitish, fibrillose or scaly. Often curved, tapering downwards and somewhat rooting.
Flesh: White, firm and thick at disc. Usually becoming reddish when broken. Strong smell of cheap soap (hence the name) and a bitter taste.
Habitat and season: Common and gregarious in all kinds of woodland. From August to November.
Edibility: Worthless.

* **Tricholoma gambosum** (Fries) Kummer
Tricholoma georgii (Fries) Quélet
St. George's mushroom

Cap: 5–15 cm. Convex to expanded, smooth and often irregular and cracked. Margin incurved and woolly at first. Light buff in colour.
Gills: Sinuate or adnexed with a tooth. Crowded, whitish and rounded behind.
Spores: White and elliptical. 5–6 × 3–3·5 μm.
Stipe: Thick, solid and curved. Whitish.
Flesh: Thick and white. Smells of new meal.
Habitat and season: Grassy places especially base rich soil growing in groups or rings. Locally common. April to July. Not recorded in N. America.
Edibility: Excellent.

95 Tricholoma atrosquamosum (Chevallier) Saccardo
Tricholoma terreum variety *atrosquamosum* (Chevallier) Massee

Cap: 4–8 cm. Campanulate or acutely umbonate with incurved margin, then expanded and often splitting. Disc fleshy but thin elsewhere. Greyish and covered with downy bluish-grey squamules, these are black or very dark and crowded at the disc.
Gills: Broad, emarginate and rather distant, thick with serrate edges. Greyish white in colour.
Spores: White and elliptical. 5–6 × 4–5 μm.
Stipe: White or grey, fibrillose, with black dots at the apex.
Flesh: Greyish, soft and fragile. Taste agreeable, peppery odour.
Habitat and season: Fairly common in both coniferous and frondose woodland from August to November, but can occur at almost any time of the year.
Edibility: Fair.

* **Tricholoma terreum** (Fries) Kummer

Almost identical to *T. atrosquamosum* but in this species the stipe apex is white mealy with no black dots. It is found in similar habitats and is probably more common.

There are a number of allied species or varieties. e.g. *T. argyraceum* (St Amans) Gillet, and *T. gausapatum* (Fries)

93 Tricholoma caligatum

94 Tricholoma saponaceum

95 Tricholoma atrosquamosum

96 Lepista nuda

Quélet. Reference should be made to more advanced works on the subject with regard to these and other close relatives.

* **Tricholoma aurantium** (Fries) Ricken
Armillaria aurantium (Fries) Kummer
Golden armillaria

Cap: 6–10(12) cm. Convex then expanded, viscid and innately squamulose. Orange, darker at disc, margin paler and incurved.
Gills: Crowded, emarginate or free. White becoming tinged rufous.
Spores: White and elliptical with oil drop. 4–5 × 2–3 μm.
Stipe: Robust, equal or slightly attenuated upwards, stuffed then hollow. With concentrically arranged tawny-orange scaly belts up to the obsolete ring. Whitish above.
Flesh: Ample and firm but thin at cap margin. White, sometimes reddish in stipe. Odour of cucumber, taste bitter.
Habitat and season: Occasional. Gregarious or in rings under pines. From September to November.
Edibility: Worthless.

* **Tricholoma ustale** (Fries) Kummer
Burnt tricholoma

Cap: 4–9 cm. Convex then plane, margin incurved, smooth, viscid in moist weather. Reddish bay-brown becoming very dark brown with age.
Gills: Crowded and rather broad, emarginate with a

decurrent tooth. White then spotted rufous.
Spores: White and elliptical with oil drop. 5–6 × 3–4 μm.
Stipe: Slender, fibrillose and equal. Stuffed then hollow, rooting. Whitish or with rufescent tinge, apex silky white.
Flesh: White and ample. No smell of meal on breaking.
Habitat and season: Fairly common in frondose woodland and rarely under conifers. From August to November.
Edibility: Not known.

* **Tricholoma ustaloides** Romagnesi

This is similar to *T. ustale* but has a strong mealy odour.
Spores: 5–7 × 3·5–5 μm.

* **Tricholoma portentosum** (Fries) Quélet
Dingy agaric

Cap: 6–10(14) cm. Campanulate or convex then expanded and often irregular. Viscid in moist atmosphere. Greyish-brown often tinged purplish, with radial blackish or purplish-black fibrils, dense at disc.
Gills: Sub-distant, very broad and thick, rounded behind, sinuate. White then greyish or yellowish.
Spores: White, smooth and elliptical. 5–6 × 3·5–4·5 μm.
Stipe: Robust and fibrillosely striate, furfuraceous above. Somewhat rooting. White and often tinged grey or yellow.
Flesh: Firm, fragile and white. Odour and taste indistinctive.
Habitat and season: Occasional, solitary or gregarious, mainly under pines. From August to November.
Edibility: Good.

* **Tricholoma populinum** J. Lange
Sandy tricholoma

Cap: 7–12 cm. Convex to plane and viscid. Light brown becoming darker with age.
Gills: Crowded and emarginate. White becoming rufous.
Spores: White and elliptical with oil drop. 5–6 × 3–4 μm.
Stipe: Robust and fibrous with a bulbous base. White becoming pale rufous.
Flesh: Clear white. Odour and taste strong of meal.
Habitat and season: In grass under poplars especially in sandy places. From September to November.
Edibility: Of poor quality but not poisonous.

* **Tricholoma sulphureum** (Fries) Kummer
Tricholoma bufonium (Fries) Gillet
Sulphur tricholoma

Cap: 4–8 cm. Convex then plane or depressed, irregular and undulate, not viscid. Silky then smooth. Sulphur-yellow, centre may be tinged brownish.
Gills: Distant, adnexed and cut out behind. Sulphur-yellow.
Spores: White and elliptical. 8–11 × 5–6 μm.
Stipe: Long and fibrillose, often curved, stuffed then hollow. Sulphur-yellow.
Flesh: Fragile and yellow. Has a strong odour of gas tar and a mild taste.
Habitat and season: Common in mixed woods especially oak. August to November.
Edibility: Worthless and suspect.

* **Tricholoma columbetta** (Fries) Kummer
Dove-coloured tricholoma

Cap: 6–10 cm. Convex or campanulate then plane. Minutely tomentose then silky fibrillose or smooth. Margin thin and at first involute. Shining white then spotted blue, greenish, violaceous or pink.
Gills: Rather crowded, broad, thin and sinuate then nearly free. Edge slightly serrate. White.
Spores: White, smooth and elliptical. 5–7 × 4–5 μm.
Stipe: White, stout, fibrillose and rooting.
Flesh: Firm and not thick. Taste and odour indistinctive.
Habitat and season: Occasional under frondose trees especially beech. Either solitary or gregarious. August to November.
Edibility: Good.

* **Tricholoma fulvum** (Fries) Saccardo
Tricholoma flavobrunneum Fries

Cap: 5–8(13) cm. Conically convex then expanded with umbo. Viscid and radially streaked with innate fibrils, often wrinkled. Rich brown, darker at disc when moist.
Gills: Crowded and emarginate with a decurrent tooth. Pale yellow becoming brown-spotted.
Spores: White and elliptical with an oil drop. 5–7 × 3–4·5 μm.
Stipe: Rather long, slender, hollow and fibrillose. Usually attenuated at both ends, viscid when moist. Concolorous with cap but paler at apex.
Flesh: White in cap, distinctly yellow in stipe. Taste and odour strong of rancid meal.
Habitat and season: Fairly common, often caespitose, on heaths and under frondose trees, especially birch. From September to November.
Edibility: Poor quality but not poisonous.
The yellow flesh of the hollow stipe is characteristic.

* **Tricholoma flavovirens** (Fries) Lundell
Tricholoma equestre (Fries) Quélet
Yellow knight fungus

Cap: 5–10 cm. Convex then plane and often undulate, smooth, viscid. Sulphur or greenish-yellow with minute reddish squamules at disc. Cuticle easily removed.
Gills: Crowded, broad, ventricose and free or barely attached by a tooth. Bright sulphur-yellow.
Spores: White, smooth and elliptical. 5·5–8 × 4–5 μm.
Stipe: Thick, short, firm and fibrillose. Pale sulphur-yellow with rusty or olivaceous zones.

Flesh: White and firm. Thick at disc and tinged yellow under cuticle. Taste and odour agreeable (mealy).
Habitat and season: Fairly common under pines especially in sandy areas. Rare under frondose trees. From July to November.
Edibility: Very good.

* **Tricholoma albobrunneum** (Fries) Kummer

Cap: 5–8 cm. Campanulate then expanded, fleshy at disc. Viscid and streaked with innate fibrils, often wrinkled. Margin incurved at first. Chestnut-brown.
Gills: Scarcely crowded, very broad, emarginate and rounded behind. White, becoming darker with age.
Spores: White and elliptical, with oil drop. 4·5–5 × 3–4 μm.
Stipe: Shortish and more or less equal. Concolorous with cap but white at base and white mealy at apex.
Flesh: Firm and ample, clear white. Odour and taste indistinctive or mild of meal.
Habitat and season: Common and often growing in rings under pines. From September to November.
Edibility: Of poor quality and indigestible.

* **Tricholoma vaccinum** (Fries) Kummer
Scaly tricholoma

Cap: 4–8 cm. Umbonato-convex then plane with umbo, dry. Margin involute, appendiculate and shaggy. Russet or reddish-brown and covered with concolorous, plushy scales.
Gills: Sub-distant, broad, thick and sinuate. Pallid then russet or spotted russet.
Spores: White, smooth and elliptical. 5·5–7 × 4–4·5 μm.
Stipe: Slender and equal, or slightly attenuated towards apex, hollow. Remarkably reddish-brown, fibrillose apart from apex which is pruinose and whitish tinged reddish-brown.
Flesh: Thick, fibrous and white, later reddish. Taste unpleasant.
Habitat and season: Fairly common, tufted or in groups under pines. From August to November.
Edibility: Worthless.

* **Tricholoma imbricatum** (Fries) Kummer

Similar to *T. vaccinum* although usually darker, but differs in the lack of appendiculate shaggy margin and the white stipe which is powdery or pruinose throughout.
Spores: White and elliptical. 5–9 × 4–5·5 μm.

96 Lepista nuda (Fries) Cooke
Tricholoma nudum (Fries) Kummer
Wood blewit or Naked mushroom

Cap: 6–10 cm. Convex, then expanded and finally depressed. Young specimens have incurved margin. Purple violet, darker in centre. Becoming dingy and brownish with age. Moist and smooth.
Gills: Crowded, narrow and sinuate. Pale violet but changing colour with age and becoming brownish-violet.
Spores: Pale-pink and minutely prickly. 6–8 × 4·5–5 μm.
Stipe: Solid and elastic, often somewhat bulbous at base. Violet or greyish-violet with lighter silky fibrils.
Flesh: Bluish-lilac when moist, whitish when dry. The odour is aromatic and pleasant to some, others find it disagreeable.
Habitat and season: Gregarious and common under frondose or coniferous trees, from September to early winter. It can withstand slight frosts. An easily identifiable species.
Edibility: Good. It should be cooked slowly and the surplus liquid (which exudes copiously) poured off.

* **Lepista saeva** (Fries) P. D. Orton
Rhodopaxillus saevus (Fries) Maire
Tricholoma saevum (Fries) Gillet
Tricholoma personatum (Fries) Kummer
The Blue-leg or Field blewit

Similar to *L. nuda*, usually larger and not quite so vivid in its violet tints, gills never violet. The cap is often clay-coloured. Found September to December mainly in pastures, gregariously or in rings. Edible and good.

97 Lepista sordida (Fries) Singer
Tricholoma sordidum (Fries) Kummer
Dirty blewit

Cap: 3–7 cm. Convex, slightly umbonate and smooth with incurved margin. Later flattened or depressed. Livid lilac at first becoming violet fuscous and finally dirty fuscous.
Gills: Sinuate to slightly decurrent. Rather crowded becoming less so with age. Violaceous becoming pale and dingy.
Spores: Rough and elliptical. Pale greyish-lilac in the mass. 6–7 × 3·5–4·5 μm.
Stipe: Pliant and fibrilloso-striate. Thickened at base. Of a similar colour to cap.
Flesh: Pale violet.
Habitat and season: Common on lawn mowings, dung-heaps, under growing crops such as potatoes etc. Summer to late autumn.
Edibility: Yes.

* **Lepista luscina** (Fries) Singer
Clitocybe luscina (Fries) Karsten
Tricholoma panaeolum (Fries) Quélet

Cap: 5–10 cm. Convex then plane or often depressed, usually wavy and eccentric. Sooty grey or brown, pruinose, becoming paler and often spotted. Margin involute, white and mealy.
Gills: Very crowded and sinuate, at length slightly decurrent. Dirty white then greyish-violet, often with a rufous tinge.
Spores: Pinkish in the mass, rough and elliptical. 5–6 × 3·5 μm.
Stipe: Thick, equal and glabrous, not polished. Fibrous outside, spongy within. Greyish-white to brownish.
Flesh: White and spongy. Odour of meal.
Habitat and season: Common and gregarious in pastures. September to November.
Edibility: Good.

* **Lepista irina** (Fries) Bigelow
Tricholoma irinium (Fries) Kummer
Rhodopaxillus irinus (Fries) Métrod

Cap: 5–10 cm. Convex then plane, smooth with wavy margin. Pale clay colour to greyish.
Gills: Crowded, narrow and sinuate. Flesh colour.
Spores: Pinkish in the mass and minutely prickly. 7–8 × 3·5–4 μm.
Stipe: Thick and fibrillar, pallid and pruinose. Often tinged ochraceous at base.
Flesh: Thick and white with the characteristic odour of violets or orris root.
Habitat and season: Fairly common and gregarious on the ground in wood-clearings or shady pastures. September to November.
Edibility: Good but slightly purgative.

Genus-*Tricholomopsis*

As Tricholoma, but gill-edged with large thin-walled pear-shaped cystidia.

98 Tricholomopsis rutilans (Fries) Singer
Tricholoma rutilans (Fries) Kummer
Purple blewit

Cap: 5–15 cm. Campanulate or convex then expanded. Ground colour yellow but covered with dense downy purplish scales which are thicker at the disc.
Gills: Chrome-yellow, crowded and broad. Gill-edge thick and slightly woolly.
Spores: White and broadly elliptical. 5–7 × 4–5 μm.
Stipe: Equal or tapering downwards. Pale yellow and covered with purple scales like the cap.
Flesh: Pale yellow, soft and thick.
Habitat and season: Common and gregarious on and around conifer stumps. From August to November. A very handsome species readily identified by its yellow gills and purple scaly cap.
Edibility: Indigestible and of poor quality.

99 Tricholomopsis decora (Fries) Singer
Clitocybe decora (Fries) Gillet
Tricholoma decorum (Fries) Quélet
Pleurotus decorus (Fries) Saccardo

Very similar to *Tricholomopsis rutilans* but with brown scales on yellow cap. Found on and around conifer stumps in northern areas. From August to November.

100 Tricholomopsis platyphylla (Fries) Singer
Collybia platyphylla (Fries) Kummer
Broad-gilled tricholoma

Cap: 4–12 cm. Convex at first then expanded. Brownish-grey and streaked radially with brown fibrils.
Gills: Adnate or adnexed. White, very broad and becoming distant in old specimens.
Spores: White. 6·5–8·5 × 6–7 μm.
Stipe: Greyish, stout, fibrous, equal and striate. Rooted on substrate by thick white mycelial strands (see illustration).
Flesh: White and thin, especially at cap margin.
Habitat and season: Found fairly commonly in frondose woods growing on rotten wood or on the ground when the mycelial strands will be attached to buried twigs, dead roots or leaves. May to November.
Edibility: Quite tasteless.

97 Lepista sordida

98 Tricholomopsis rutilans

99 Tricholomopsis decora

100 Tricholomopsis platyphylla

Genus-Armillaria

The fruit-body has a smooth or scaly cap centrally stipitate. The flesh in cap and stipe is continuous. The gills are adnate or slightly adnato-decurrent. The spores are white. There is a membranous ring present but no volva.

101 Armillaria mellea (Fries) Kummer
Armillariella mellea (Fries) Karsten
Honey fungus or Boot-lace fungus

Cap: 3–10(14) cm. Convex then expanded, often becoming centrally depressed, margin striate. Very variable in colour, yellowish, olive, tawny or sooty brown, sometimes dark grey or even pinkish. When young, it is downy or flecked with dark coloured scales, especially at the disc. With age the marginal scales disappear.
Gills: Sub-distant, adnate or slightly decurrent. Whitish then pale brownish-yellow; older specimens are brown spotted.
Spores: Whitish in the mass and elliptical 8–9 × 5–6 μm. Very copious, they can often be seen on the caps of the fungi or on the substrate from which they are growing and even on nearby herbage.
Stipe: Externally rigid and more or less grooved. Equal or swollen at the base. Concolorous with cap at first but whitish above ring, becoming reddish-brown with age and always yellowish-olive at base.
Ring: Whitish, flecked at margin with yellow and sometimes in addition with reddish-brown.
Flesh: Spongy. White, yellowish or pinkish in the cap. Brown and fibrous in the stipe. Odour strong and disagreeable, taste bitter and lingering.
Habitat and season: This conspicuously polymorphic species is commonly found in dense clusters, often tiered on and around the base of old trees, shrubs and stumps, also singly or in small clusters away from visible wood but always attached thereto by thick blackish rhizomorphs (boot-laces). These rhizomorphs can often be found under the loose bark of affected trees. The active mycelium is luminous and can be quite disturbing to human nocturnal ramblers. It is both saprophytic and parasitic. The appearance of this fungus anywhere is a sure sign of sickness in the host/s and it is not conservative in this aspect, attacking most trees and shrubs, especially old, weakened or shocked specimens. These eventually perish but the process can take many years; an old hawthorn known to the writer still thrives after fifteen years of bearing fruit-bodies of this species. Up to present there is no known cure for affected hosts.
Edibility: Eaten by many, but even after long cooking and pouring off the liquid, if often retains its bitterish taste.

N.B. There are many closely related species which may account for the so-called polymorphic nature of this fungus. *Armillaria tabescens* differs for instance in the lack of a ring on the stipe.

102 Armillaria albolanaripes Atkinson

Cap: 5–12 cm. Convex then plane, sometimes with obscure umbo. Margin viscid, involute at first and covered with ragged pure white veil remnants. Bright yellow to mustard-yellow with reddish to cinnamon flattened scales at the disc; these become darker with age.
Gills: Adnexed, not crowded and broad. White becoming cream or yellowish with age. Edges dentate.
Spores: White, smooth and elliptical. 5–7 × 3–4·5 μm.
Stipe: White to yellowish. Densely covered with shiny white floccules below the ring zone.
Ring: White, superior and lax. Short and ragged.
Flesh: Pale and spongy.
Habitat and season: Solitary or gregarious under trees or on roadside verges. August to October. Not recorded in Britain.
Edibility: Not known.

Genus-Clitocybe

Contains medium to large species which are usually terrestrial. The cap is infundibuliform and fleshy, regular or not, with margin incurved at first. The gills are more or less decurrent. The spores are white. The stipe is central and fleshy, the flesh being continuous with that of the cap.
Some are considered good esculents, whilst a few are undoubtedly *very poisonous*, even *deadly*.

103 Clitocybe flaccida (Fries) Kummer
Lepista inversa (Fries) Patouillard

Cap: 5–8 cm across. Orange brown often with a darker centre. Funnel-shaped, smooth and moist. Margin involute. The cap is paler when dry.
Gills: Thin and narrow, decurrent on the stem. Whitish at first, later becoming a palish orange.
Spores: White and sub-globose. 4–4·5 × 3–4 μm.
Stipe: Roughly the colour of the cap, short and firm. Woolly at base.
Flesh: Tough, thin and fawn in colour. Smell is slightly anise.
Habitat and season: Found growing in groups from September to December. It shows a preference for coniferous woods.
Edibility: Of poor quality, not recommended.

* **Clitocybe odora** (Fries) Kummer
Clitocybe trogii (Fries) Saccardo

Cap: 4–8 cm. Convex then flattened or depressed with a slight umbo. Margin irregular and incurved. Bluish-green and pubescent.
Gills: Subdistant and narrow, adnato-decurrent. Duller and paler than cap.
Spores: White and elliptical. 6–7·5 × 3–4 μm.
Stipe: Bluish-green and often curved. Base downy white.

101 **Armillaria mellea**

102 **Armillaria albolanaripes**

103 **Clitocybe flaccida**

104 **Clitocybe suaveolens**

Flesh: White to greenish with a strong odour of anise.
Habitat and season: Common in leaf litter growing singly or caespitose, chiefly in frondose woods. From August to November.
Edibility: May be eaten but very aromatic, better used as flavouring.

★ **Clitocybe fragrans** (Fries) Kummer

Cap: 2–4 cm. Convex and depressed, hygrophanous. Yellowish-grey when wet, whitish when dry.

Gills: Tinged cap colour.
Spores: White, long and pip-shaped. 7–8 × 3·5–4 μm.
Stipe: Equal and slightly curved. Similar in colour to cap.
Flesh: Thin and whitish with a distinct odour of anise.
Habitat and season: Occasional amongst leaves, grass and moss under frondose trees. August to December.

104 Clitocybe suaveolens (Fries) Kummer

A very similar species but found under pines. It also smells of anise.

105 Clitocybe dilatata Persoon ex Karsten

Although originally discovered from Europe, this is thought to be a frequent species in western N. America. It remains to be seen what this species really is and whether it is simply a form of *Clitocybe cerussata* (Fries) Kummer, as European authors consider. The latter is widespread in the British Isles and distinguished by its metallic white cap.

*** Clitocybe cerussata** (Fries) Kummer POISONOUS
White-lead clitocybe

Cap: 4–8 cm. Convex then plane, fleshy, minutely floccose then more or less glabrous. Margin downy and involute. White, as if painted with white-lead paint and remaining so in age.
Gills: Very crowded, adnate then decurrent. Narrow, thin and white.
Spores: White and elliptical. 5–6 × 3–4 μm.
Stipe: Smooth, tough and elastic, spongy and solid. Base often thickened and covered with white down, naked above.
Flesh: Thick at disc, thin at margin. White, soft and compact.
Habitat and season: Fairly common in frondose and coniferous woods.
Edibility: *Poisonous.*

*** Clitocybe phyllophila** (Fries) Kummer POISONOUS

Similar to *C. cerussata* but more slender. Gills broader and less crowded. Stipe becoming hollow.
Edibility: *Poisonous.*

106 Clitocybe clavipes (Fries) Kummer
Club-foot clitocybe

Cap: 4–7 cm. Convex with umbo, then flat and finally slightly depressed. Fleshy and smooth, viscid in damp weather, fibrilar or velvety when dry. Cap centre deep brownish-grey, sometimes with olivaceous tint. Paling towards the margin which is often white.
Gills: Not crowded, deeply decurrent and flaccid. White, becoming yellowish-white with age.
Spores: White and elliptical. 4·5–5 × 3·5–4 μm.
Stipe: Base very swollen and with white fibrils. Narrowing upwards. Concolorous with, or paler than cap. Spongy within.
Flesh: White, thin at cap margin. Tender, soon flaccid and watery.
Habitat and season: Common in all types of woodland especially beech and conifer. Usually in groups. From September to November.
Edibility: No.

*** Clitocybe nebularis** (Fries) Kummer

Cap: 6–15 cm. Convex and then flattened, finally slightly centrally depressed. Smooth, fleshy and often uneven. At first pale grey and with a bloom.
Gills: Crowded, thin and slightly decurrent. Whitish at first, yellowish later.
Spores: White and elliptical. 5·5–8 × 3–4·5 μm.
Stipe: Short or long, stout and tapering upwards. Fibrillosely striate, often distorted. Stuffed but later hollow.
Flesh: Thick and white with a sweetish smell.
Habitat and season: Found under frondose or coniferous trees. Locally very common, sometimes occurring in large numbers. August to December.
Edibility: Considered good but can cause indigestion in some people.

N.B. On rare occasions is parasitised by the agaric:
Volvariella surrecta (Knapp) Singer
Volvaria loveiana (Berkeley) Gillet

*** Clitocybe rivulosa** (Fries) Kummer VERY POISONOUS

Cap: 2–4 cm. Flattened to slightly depressed. Margin powdery and incurved at first. Greyish to yellowish, often wrinkled or streaked with wavy lines, more or less zoned.
Gills: Crowded, broad and slightly decurrent. Paler than cap.
Spores: White and elliptical. 3·5–5 × 2·5–3·5 μm.
Stipe: Shortish, downy at apex and concolorous with cap.
Flesh: White and spongy.
Habitat and season: Common in grass, gregarious, also in rings and often in company with *Marasmius oreades*. From August to November.
Edibility: *Very poisonous.*

107 Leucopaxillus amarus (Fries) Kühner
Clitocybe amara (Fries) Kummer
Lepista amara (Fries) Patouillard
Tricholoma amarum (Fries) Rea

Cap: 4–7 cm. Convex then expanded, obtuse or slightly umbonate. Dry and fleshy at disc. Very variable in colour ranging from rufous brown, yellowish-tawny to whitish-rufous. Slightly flocculose, often distorted, cracked and wavy.
Gills: Slightly decurrent, crowded and white.
Spores: White and elliptical, 4 × 2 μm.
Stipe: Solid, ample and tough. White floccose then nearly smooth.
Flesh: White and firm. Very bitter taste.
Habitat and season: Mixed woods from August to November.
Edibility: Not known. Suspicious.

Genus-*Laccaria*

Small species. The cap is convex, the gills adnate by a decurrent tooth and powdered with white spores. The

105 Clitocybe dilatata

106 Clitocybe clavipes

107 Leucopaxillus amarus

108 Laccaria amethystea

spores are globose or nearly so and non-amyloid. Centrally stipitate. The flesh of the cap and stipe are continuous. It grows under trees, on heathland and on bare ground.

108 Laccaria amethystea (Mérat) Murrill
Clitocybe amethystea (Mérat) Saccardo
Laccaria amethystina (Huds.) Cooke
Amethyst laccaria or Red cabbage fungus

Cap: 1–4 cm. Convex to flat and uneven. Centre often slightly scurfy especially in older specimens. Deep violet when moist but paler in dry conditions. Hygrophanous.

Gills: Adnate or slightly decurrent. Distant with intermediates, broad and thick. Colour as the cap.

Spores: White and globose with spines, 9–11 μm. Basidia 4-spored.

Stipe: Slender, elastic and tough, becoming hollow. Violet but paler in the lower half and hairy at base.

Flesh: Concolorous with rest of plant. Fibrous in stipe.

Habitat and season: Grows singly or in small groups on the ground in mixed woods, especially in damp shady places. Common August to December.

Edibility: Considered fair, if enough can be collected for a meal.

109 Laccaria laccata (Fries) Cooke
Clitocybe laccata (Fries) Kummer
Deceiver

Cap: 2–6 cm. Convex then expanded, sometimes umbilicate and margin often wavy. Hygrophanous, rich brown or brick-red in moist conditions; paler when dry and minutely squamulose.
Gills: Thick, distant and adnate by a tooth. Pinkish then powdered white.
Spores: White, globose and spiny. 7–10 μm. Basidia 4-spored.
Stipe: Usually longish, equal, fibrous and tough, often twisted. Cap-colour, hispid white at base.
Flesh: Cap-colour and thin.
Habitat and season: A very common and variable species, usually growing in troops under trees and on heaths. July to December.
Edibility: Yes.

Genus-*Omphalina*

Very small species. The cap is convex or umbilicato-depressed, the gills are decurrent. The spores are white and non-amyloid. Stipe central, thin, smooth and cartilaginous. The flesh of the cap and stipe is not homogeneous.

110 Omphalina ericetorum (Fries) M. Lange
Omphalina umbellifera (Fries) Quélet

Cap: 1–2 cm. Convex-umbilicate, then expanded and centrally depressed. Smooth, shining and radially striate. At first the margin is inflexed and crenate. Colour is very variable: white, yellow, orange or light brownish-olive but usually darker at disc.
Gills: Decurrent, thick and very distant, broad behind. White to pale yellow.
Spores: White and pip-shaped. 8–9 × 4·5–5·5 μm.
Stipe: Solid and internally cottony, tough and smooth. Attenuated downwards, downy at base. Concolorous with cap.
Flesh: Very thin and paler than cap.
Habitat and season: Common in swamps, on heaths, and often in large troops especially on peaty moors. April to November.
Edibility: Worthless.

111 Xeromphalina cauticinalis (Fries) Kühner & Maire
Marasmius cauticinalis Fries

Cap: 1–1·5 cm. Campanulately convex, obtuse and often centrally depressed, glabrous. Even, then coarsely striate or grooved, often with a jagged margin. Tawny, darker at centre.
Gills: White and elliptical. 7 × 3·5 μm. Amyloid.
Stipe: Flexuous, equal and hollow but containing a pith. Dark red-brown and tomentose.
Flesh: Very thin and yellowish.
Habitat and season: Clustered on logs and stumps, sometimes on pine needles. August to October.
Edibility: Worthless.

Genus-*Collybia*

Contains small to large species. The cap is usually convex with margin involute at first. The gills are adnate, adnexed or becoming free, usually crowded. The spores are white, smooth and non-amyloid; the stipe is central and cartilaginous.

They are found on the ground, on wood or among decaying gill fungi, often deeply rooting. Some are edible and probably none is poisonous.

112 Collybia maculata (Fries) Quélet
Spotted collybia or Rust spot fungus

Cap: 4–15 cm. Convex with strongly incurved edge, then expanded. It is tough and fleshy. At first entirely white (immaculate) but soon rust-spotted.
Gills: White to cream, either free or adnate. Narrow and very crowded. Later rust-spotted.
Spores: White or with a slightly rosy tinge. Sub-globose. 4–5·5 × 3·5–5 μm.
Stipe: Longish, firm and cartilaginous, equal or swollen. Furrowed longitudinally. White but later rust-spotted.
Flesh: White, tough and thick.
Habitat and season: Common in all types of woodland especially under bracken and usually growing in clusters. From July to November.
Edibility: Listed by some authorities as edible. It has an agreeable odour but a bitter taste.

* **Collybia erythropus** (Fries) Kummer
Marasmius erythropus (Fries) Fries
Marasmius bresadolae Kühner & Romagnesi

Cap: 2–2·5 cm. Convex then plane, obtuse, hygrophanous and striate. Pallid wrinkled and nearly white when dry.
Gills: Broad, distant and almost free, connected by veins. Whitish to pale flesh colour.
Spores: White. 6–9 × 4–5 μm.
Stipe: Hollow and firm, smooth and shining. Round or compressed. Blackish-red, becoming paler at apex, white strigose at base.
Flesh: Thin and white in cap, reddish-brown in stipe. Taste mild.
Habitat and season: Common on the ground in frondose woods. September to November.
Edibility: Good.

109 Laccaria laccata

110 Omphalina ericetorum

111 Xeromphalina cauticinalis

112 Collybia maculata

113 Collybia peronata (Fries) Kummer
Marasmius peronata (Fries) Fries
Wood woolly-foot

Cap: 2–5 cm. Convex, then expanded. Thin and wrinkled, reddish to purplish-brown. Margin striate.
Gills: Adnexed, distant or crowded. Thin and pallid but later rufous.
Spores: White, pip-shaped to lanceolate. 7·5–9 × 3–3 μm.
Stipe: Pliant and stuffed with fibres. Yellow to brownish-yellow and covered with yellowish woolly hairs at base.
Flesh: Ample and leathery. Yellowish-brown.
Habitat and season: Grows gregariously, mainly in the leaf litter of frondose woods. Very common from August to November.
Edibility: Worthless. Tastes peppery when chewed.

* **Collybia fuscopurpurea** (Fries) Kummer
Marasmius fuscopurpurea (Fries) Fries

Found in similar situations as *C. peronata*. The stipe is strigose at base. Has a mild taste, not peppery. Less common. Found September to November.

114 Collybia acervata (Fries) Kummer
Marasmius acervatus (Fries) Karsten

Cap: 2–5 cm. Convex then expanded, smooth and hygrophanous. Margin slightly striate. May be of light or dark yellowish-tan, sometimes reddish.
Gills: Crowded, narrow and free. Whitish in colour.
Spores: White. 6–7 × 2·5–3 μm.
Stipe: Cap colour above; reddish-brown below. Smooth, tough and pliable. Hairy at base.
Flesh: Thin and white, odour agreeable, taste mild.
Habitat and season: Found at the base of pine-stumps. Common from May to November.
Edibility: Yes.

115 Collybia dryophila (Fries) Kummer
Marasmius dryophilus (Fries) Karsten

Cap: 2–6 cm. Convex but soon plane and often centrally depressed. Smooth, fleshy, pale to dark tan, tinged brown at disc.
Gills: Crowded, narrow and adnexed or free. White or pale yellow.
Spores: White and elliptical. 4·6–6·5 × 3–3·5 μm.
Stipe: Smooth and slender with swollen rooting base. Cartilaginous, hollow and more or less concolorous with cap.
Flesh: Thin, white and inodorous with a mild flavour.
Habitat and season: Solitary or loosely gregarious on the ground among fallen leaves, especially under oak trees. A very common and variable species especially in N. America. Found May to November.
Edibility: Uninteresting but not poisonous.

* **Collybia confluens** (Fries) Kummer
Marasmius confluens (Fries) Karsten

Cap: 2–4 cm. Deeply convex at first then flattened, thin. May be either flesh-colour, pale clay or almost white.
Gills: Very crowded and noticeably narrow, adnexed or free. Concolorous with cap.
Spores: White and elliptic. 6–7(9) × 3–4 μm.
Stipe: Confluent, downy, slender and tough, often compressed. Concolorous with cap or with a reddish-purple tinge below.
Flesh: Pallid in cap; brownish in stipe, with strong musty odour.
Habitat and season: Grows densely clustered, often in rings, amongst leaf debris in woods, especially beech. Very common from July to August.
Edibility: Yes.

Genus-*Melanoleuca*

Medium to large hygrophanous species. Similar to Tricholoma but with some different microscopical characteristics which include rough, amyloid spores.

116 Melanoleuca grammopodia (Fries) Patouillard
Tricholoma grammopodium (Fries) Quélet
Ring agaric

Cap: 6–8 cm. Campanulate, then plane with umbo. Margin incurved until adult. Brownish-grey, smooth and hygrophanous.
Gills: Curved, crowded and adnate. White turning grey.
Spores: White. 8–10 × 5–6 μm.
Stipe: White with brownish-grey striations. Long, equal and stuffed at first with silky flesh. Later hollow. The bulbous base is covered with a white cottony substance (see illustration).
Flesh: Spongy and whitish with a disagreeable odour.
Habitat and season: Clearings in woodlands, also in pastures, especially where grass is lush. Gregarious and forming rings. From August to November.
Edibility: Not recommended because of the nasty odour which it retains even when cooked.

* **Melanoleuca melaleuca** (Fries) Murrill
Melanoleuca vulgaris Patouillard

Somewhat similar to *M. grammopodia* but smaller. This fungus shows a great variation in colour due to changes in water content.

Genera-*Panus* and *Panellus*

Medium to small species with short lateral stipe and growing on wood. The gills are fan-like or decurrent, with an even edge not serrated.

113 Collybia peronata

114 Collybia acervata

115 Collybia dryophila

116 Melanoleuca grammopodia

The spores are white, cylindric and smooth. In Panus non-amyloid and Panellus amyloid.

117 Panellus stipticus (Fries) Karsten
Panus stipticus Fries
Styptic toadstool

Cap: 1–4 cm convex to flat with scalloped edge. Pale tan to cinnamon and roughly kidney-shaped. Resembles an ear when held by the stipe. Thin but elastic in consistency.
Gills: Narrow, crowded and concolorous with cap or may have a yellowish tinge.
Spores: White. 3·5–4·5 × 2–2·5 μm. Amyloid.
Flesh: Tough.
Habitat and season: The overlapping ear-shaped fruit-bodies can be found in clusters on tree stumps, chiefly hardwoods, from January to December.
Edibility: This fungus is reputed to be poisonous and is very bitter to the palate. If a small fragment is chewed, its astringency is at once evident.

Slugs regularly make inroads into the clusters, often clearing the entire growth from a stump.
In America, this species is known to be luminous, particularly the gills, but also the cap and the mycelium. The British form, as far as is known, is non-luminous.
The illustration also shows the underside of the myxomycete *Didymium squamulosum* (A. & S.) Fries.

* **Panellus serotinus** (Fries) Kühner
Pleurotus serotinus (Fries) Kummer

Cap: 4–10 cm. Convex and gibbous then plane, margin incurved. Viscid in wet weather. Colour variable, yellowish-green, brownish-green or olive green.
Gills: Crowded and narrow, often branched. Yellow then darker.
Spores: White, long-cylindric and smooth. 4–5·5 × 1–2 μm. Amyloid.
Stipe: Lateral, short or almost wanting. Yellowish with olivaceous or brown squamules especially near gills.
Flesh: Thick, white and gelatinous under cap cuticle.
Habitat and season: Occasional on the trunks of frondose trees. From September to November.
Edibility: Worthless.

Genus-*Pleurotus*

Medium to large species. The cap is fleshy and bracket-like. The gills are decurrent, sinuate or adnate and often fan-like. Spores are white, smooth and sausage-shaped, non-amyloid. Grows mainly on wood.
Some are good to eat when young.

118 Pleurotus ostreatus (Fries) Kummer
Illustration shows variety *columbinus* (Quélet) Quélet
Oyster mushroom

Cap: 3–15 cm. Shell-shaped and smooth with incurved margin. Later expanded. Colour very variable: dark violaceous, brownish or greyish-ochre when young but fading and becoming dingy with age.
Gills: Crowded, broad and decurrent. White and then dirty white.
Spores: White with lilac tinge. Sausage-shaped, 10–11 × 3·5–4 μm. Non-amyloid.
Stipe: Eccentric, short or virtually absent. Solid and branching.
Flesh: Thick and tender when young. Becoming tough and rubbery with age. White.
Habitat and season: Grows clustered in overlapping tiers from a common base on the trunks of living or dead trees, mainly frondose and especially beech. Common at any time of the year except during periods of hard frost.
Edibility: Good but should be cooked slowly. It is grown commercially in some countries.

119 Pleurotus cornucopiae (Persoon) Rolland
Pleurotus ostreatus variety *cornucopiae* (Persoon) Quélet
Oyster mushroom

Cap: 5–12 cm. Convex, later expanded and depressed. Downy at first but soon glabrous. Off-white in colour, frequently tinged pink, then brownish or ochre.
Gills: Crowded at cap margin with intermediates, but less so near stipe on which they are decurrent and often form a net-like pattern.
Spores: White. 8–11 × 3·5–5 μm.
Stipe: Usually eccentric, curved and branched. White or yellowish, finally brownish.
Flesh: Thick, tender and soft. White and smells of rancid flour.
Habitat and season: Fairly common growing clustered on old trunks of hardwood trees especially elm and oak. April to October.
Edibility: Good.

120 Panus torulosus (Fries) Fries

Caps: 4–10 cm across. Fan-shaped or irregularly funnel-shaped with the margin curled into lobes and fissured towards the gills. Yellowish flesh-colour or pale tan. Fleshy and pliant becoming tough.
Gills: Pale tan tinged flesh-colour to violaceous. Narrow, not crowded and deeply decurrent.
Spores: White, smooth and oblong-elliptical. 5–6 × 3 μm.
Stipe: Eccentric to lateral. Short and stout. Pale tan, but when young covered with a violet down which soon disappears.
Flesh: White and firm.
Habitat and season: Occasional on old stumps and fallen trees. Usually in clusters. From July to November.
Edibility: No.

117 Panellus stipticus

118 Pleurotus ostreatus

119 Pleurotus cornucopiae

120 Panus torulosus

121 Oudemansiella mucida (Fries) Höhnel
Armillaria mucida (Fries) Kummer
Mucidula mucida (Fries) Patouillard
Porcelain fungus

Cap: 3–8 cm across. Shiny white and often shading to olivaceous at the centre. Hemispherical, and thin at the margin. Translucent and very slimy.
Gills: White, broad and distant, also rounded towards the stipe.
Spores: White and almost round. 13–18 × 12–15 μm.
Stipe: White, slender and stiff. Usually curved.
Flesh: Thin, soft and white with a delicate odour.
Habitat and season: Usually found on dead trunks and branches or on weak trees, predominantly beech. A fairly common species from August to November.
Edibility: Has a mild flavour and said to be edible after washing.

122 Oudemansiella radicata (Fries) Singer
Collybia radicata (Fries) Quélet
Mucidula radicata (Fries) Boursier
Rooting shank

Cap: 3–9 cm. Umbonato-convex then plane and somewhat gibbous. Radially wrinkled, glutinous in moist conditions. Yellow-brown or olive-brown.
Gills: Distant, especially with age. Thick and adnexed often becoming free. Shining white at first, sometimes edged brown, dingy in old specimens.
Spores: White and broadly elliptical. 12–16 × 10–12 μm. Non-amyloid.
Stipe: Long, erect and rigid. Attenuated upwards and often twisted. It extends root-like into the ground for a considerable distance until in contact with wood. Pale brown or whitish.
Flesh: White, soft and pliable.
Habitat and season: Usually growing singly, occasionally in twos, near to and under frondose trees and stumps in open or dense woodland. Common from July to November.
Edibility: Yes.

* **Flammulina velutipes** (Fries) Karsten
Collybia velutipes (Fries) Kummer
Winter fungus or Velvet shank

Cap: 2–6 cm. Convex then plane. Fleshy at disc but thin at margin. Often eccentric and irregular. Smooth and viscid. Tawny-yellow, darker at disc.
Gills: Adnate or adnexed, sub-distant and very unequal. Yellowish becoming tawny.
Spores: White. 6·5–10 × 3–4 μm.
Stipe: Tough, cartilaginous and equal, often contorted. Yellow above to black below or may be wholly dark brown.
Flesh: Yellowish, watery and soft. Odour and taste pleasant.
Habitat and season: Caespitose on living or dead trees. Common from August to April.
Edibility: Yes, and good after discarding the stipes.

Genus-*Mycena*

The genus is readily recognisable in the field by its macroscopic characteristics. The species are mainly small or very small. The cap is more or less campanulate. When young the margin is never incurved and usually striate. The gills are ascending, less often arched or decurrent; light coloured or white pruinose with the spores. Examination of gill cystidia under a microscope is important in identification. The spores are white, amyloid or non-amyloid; smooth and usually ellipsoid; basidia 2- or 4-spored. The stipe is central and usually slender, viscid or dry, smooth or pubescent, often longitudinally furrowed. Some species exude a coloured latex when the stipe is broken and squeezed, this is a useful guide to identification. The flesh is thin. Taste is unimportant except in *Mycena erubescens* Höhnel, which is very bitter like quinine. Odour is a useful guide in identification to persons with a keen sense of smell.

They occur on living or dead trees, on the ground, or on leaf debris and twigs. As esculents they are quite worthless, but none are known to be poisonous to man.

123 Mycena alcalina (Fries) Kummer

Cap: 1–3 cm. Conic-campanulate, striate. Dull grey or grey-brown but can be olive tinted or nearly black.
Gills: Subdistant, adnate, linear or ventricose. Whitish then brownish or dark grey, edged white.
Spores: White, elliptic-cylindric. Basidia 4-spored. 8–12 × 4·5–6 μm. Amyloid.
Stipe: 5–8 cm long. Slender and rigid, smooth and shining with a villose base. More or less concolorous with cap.
Flesh: Whitish and thin. Has a mild taste and nitrous smell.
Habitat and season: Fairly common growing on and around conifer stumps. From September to November.
Edibility: Worthless.

* **Mycena leptocephala** (Fries) Gillet

Very similar to *M. alcalina* but it grows in short grass on lawns and in fields etc.
Spores: 10·5–12 × 5·5 μm. Amyloid.

* **Mycena epipterygia** (Fries) S. F. Gray

Cap: 1–3 cm. Campanulate. Covered with a viscid, elastic, separable pellicle. Greyish or with disc livid brown or yellow. Becoming pale green or creamy.
Gills: Subdistant, adnate or subdecurrent. White or pale pink, edge with a detachable glutinous thread.
Spores: White. 8–10 × 4·5–5 μm. Amyloid.

121 Oudemaneiella mucida

122 Oudemansiella radicata

123 Mycena alcalina

124 Mycena aurantiidisca

Stipe: 5–8 cm long. Slender and equal. Lemon-yellow, base white strigose.
Flesh: White and very thin, with a mild taste but odourless.
Habitat and season: Common growing singly or a few together, on heaths and in coniferous woods from September to November.
Edibility: Worthless.

124 Mycena aurantiidisca Murrill

Cap: 0·5–2 cm. Obtusely conical when young, not expanding later, but merely broadly conical. Glabrous, striate when moist. Orange when young fading to mustard-yellow at the disc and whitish at the margin.
Gills: Crowded to sub-distant and bluntly adnate. White then tinged yellowish with age. Margin concolorous with gill face.
Spores: White and ellipsoid. 7–8 × 3·5–4 μm. Non-amyloid.
Stipe: Equal, hollow and fragile. Minutely pruinose above, scarcely so at base. White at apex to yellowish below.
Flesh: Thin and fragile, orange or yellow. Odour and taste mild.
Habitat and season: Uncommon. Found under conifers during spring and autumn in Idaho, Washington, Oregon and British Columbia. Not recorded in Britain.
Edibility: Worthless.

N.B. Considered by some authorities as a form of *Mycena adonis* (Fries) S. F. Gray—a species which is found in Britain.

125 Mycena fibula (Fries) Kühner
Omphalina fibula (Fries) Kummer

Cap: 0·5–1 cm. Convex then flattened or slightly umbilicate. Smooth and striate. Light orange but darker at the disc.
Gills: White to pale yellow, arched and decurrent by a tooth.
Spores: White and narrowly elliptical. 4–5 × 2 μm. Non-amyloid.
Stipe: 2–4 cm long by 1 mm, equal. Light orange and downy. Darker below and with a hairy base.
Flesh: Orange and thin.
Habitat and season: Grows commonly, chiefly amongst moss in shady sites. A pretty fungus which is often overlooked because of its small size. From May to November.
Edibility: Worthless.

* **Mycena swartzii** (Fries) Kühner

Differs from *M. fibula* in that the cap is pale ochraceus with a blackish-brown disc. The stipe is paler and purplish-brown at apex. Flesh is concolorous with the cap. In all other respects it is similar to *M. fibula*.

126 Mycena galericulata (Fries) S. F. Gray
Mycena rugosa (Fries) Quélet
Bonnet mycena

Cap: 2–5 cm. Campanulate then expanded and broadly umbonate, striate to the umbo. Grey-brown or yellow-brown to dingy white.
Gills: More or less distant, broad, adnate and connected by veins at base. Whitish then flesh-coloured.
Spores: White. 9–12 × 6–8 μm. Amyloid. Basidia usually 2-spored.
Stipe: Long or short, tough, smooth, hollow and rooting. Concolorous with the cap.
Flesh: Whitish and thin.
Habitat and season: Grows singly or clustered on stumps and fallen trunks, mainly of frondose trees. Very common from January to December.
Edibility: Worthless.

127 Mycena galopus (Pers. ex Fries) Kummer

Cap: 1–2 cm. Campanulate and striate. Greyish to pale tan with darker disc.
Gills: White to grey. Subdistant, adnate. Closely hairy (lens).
Spores: Cartridge buff, elliptic and smooth 10–13 × 5–6 μm. Amyloid.
Stipe: 5–10 cm × 1–2 mm, equal. Smooth and same colour as cap. Woolly at base and somewhat rooting. Yields white juice when broken.
Flesh: White and very thin.
Habitat and season: Very common on the ground in woods, usually in troops attached to twigs and leaves etc. August to December.
Edibility: Mild taste and no distinctive smell, worthless.

The variety *candida* Lange. is similar but completely white.

128 Mycena leucogala (Cooke) Saccardo

Similar to *M. galopus* but:
Cap: Very dark brown or even black.
Gills: Grey.
Stipe: Grey to black.
Habitat and season: Common in burnt or peaty ground from August to November.
Edibility: Not known.

* **Mycena adonis** (Fries) S. F. Gray
Marasmiellus adonis (Fries) Singer

Cap: 0·5–2 cm. Convex or conical. Coral or orange-red, fading with age. Margin paler and striate.
Gills: Adnate, narrow and sub-distant. White then tinged pink.
Spores: White and sub-cylindrical. 9–12 × 3·5–5·5 μm. Non-amyloid. Basidia 2-spored.
Stipe: Equal and smooth. Base strigose and rooting. Whitish.
Flesh: Very thin. Red in cap, white in stipe.
Habitat and season: Occasional, often growing singly in short grass or on wood of both deciduous and coniferous types. From September to November.
Edibility: Worthless.

* **Mycena acicula** (Fries) Kummer
Marasmiellus aciculus (Fries) Singer

Cap: 2–10 mm. Orange-vermilion. Margin paler and striate.
Gills: Yellow with paler edge.
Spores: White and fusiform, 9–12 × 3–4 μm. Non-amyloid.
Stipe: Tough and bright yellow, then fading.
Habitat and season: Often growing singly attached to wood. Common from May to November.
Edibility: Worthless.

* **Mycena hiemalis** (Fries) Quélet
Marasmiellus hiemalis (Fries) Singer

Cap: 0·5–1·5 cm. Grey-brown with darker disc.
Gills: Adnate and whitish.
Spores: White and ovate. 8–9 × 5·5–6 μm. Non-amyloid. Basidia mostly 2-spored.
Stipe: Equal, whitish and densely pruinose.
Flesh: Brown to white. Thin and odourless.
Habitat and season: Grows on the mossy bark of living deciduous trees. Common from August to November.
Edibility: Worthless.

125 Mycena fibula

126 Mycena galericulata

127 Mycena galopus

128 Mycena leucogala

129 Mycena inclinata (Fries) Quélet
Mycena galericulata var *calopus* (Fries) Karsten

Cap: 2–4 cm. Conical or convex then somewhat expanded. Striate or sulcate to the disc. Shining when dry. Dingy brown or grey with paler margin which is often toothed.
Gills: Crowded, adnate and broad. White at first then pale flesh colour.
Spores: White and broadly elliptical. 8–10 × 6–7 μm. Amyloid.
Stipe: 6–10 cm. Slender and often curved upwards. Whitish above but lower part turns brown, then rich chestnut, darker towards the base.
Flesh: Whitish but brownish in the stipe.
Habitat and season: Common and very densely caespitose on stumps or buried branches of frondose trees, chiefly oak. August to November.
Edibility: Worthless.

* **Mycena tintinnabulum** (Fries) Quélet

Cap: 1–2 cm. Convex to expanded, often slimy. Cap edge not overlapping the gills. Margin slightly striate. Dark brown fading to greyish-brown.
Gills: Crowded, broad, bow-shaped and adnate with a decurrent tooth. Pale grey.
Spores: White. 4–5 × 2·5–3 μm. Amyloid.
Stipe: Short, curved and very tough. Pallid above, darker and strigose or downy below.
Flesh: Thin, whitish and very tough.
Habitat and season: Densely tufted or gregarious on stumps or fallen trunks of frondose trees. A rare species found from October to February.
Edibility: Worthless.

130 Mycena oortiana Hora
Mycena arcangeliana var. *oortiana* Kühner

Cap: 1–5 cm. Striate and obtusely conical. Grey-brown at first, later whitish with an olive tint.
Gills: Crowded and adnexed. White turning pinkish.
Spores: Slightly buff-coloured and pip-shaped. 7–8 × 4·5–5 μm. Amyloid.
Stipe: Longish, equal and polished. Dark lilaceous in colour. Base covered with downy white hairs.
Flesh: Thin. White in cap; grey in stipe. Smells strongly of iodoform.
Habitat and season: Grows tufted on stumps and branches in frondose woods. Common during the autumn. Not recorded in America.
Edibility: Worthless; has a mild taste.

131 Mycena sepia J. Lange

Cap: 1–2·5 cm. Campanulate with obtuse umbo, striate and smooth. Greyish-brown or darker, with paler margin.
Gills: Crowded, narrow and adnexed. White or grey at base.
Spores: White and pip-shaped. 8–10 × 5–6 μm. Amyloid. Basidia 2- or 4-spored.
Stipe: Equal and rather long, flaccid. Greyish-brown, base strigose and rooting.
Flesh: White. Mild taste and with an odour of iodoform.
Habitat and season: Common in deciduous woodland. September to November.
Edibility: Worthless.

132 Mycena sanguinolenta (Fries) Kummer
Small bleeding mycena

Cap: 4–20 mm. Conic-campanulate or hemispherical. Pale brownish-red with a purple cast. Striate to the umbo which is darker in shade.
Gills: Subdistant and adnate. Whitish or flesh-tinted, but edge if dark reddish-brown.
Spores: Cartridge-buff, pip-shaped. 8–10 × 4–5·5 μm. Amyloid.
Stipe: 5–8 cm × 1 mm. Reddish-brown, strigose at base. Yields blood-red juice when broken.
Flesh: Reddish and very thin with little or no smell.
Habitat and season: Very common everywhere especially in moss and particularly in the beds of needles in pine woods. It grows singly or in troops. From August to November.
Edibility: Worthless.

Also shown in the illustration is the phycomycete:
Spinellus megalocarpus (Corda) Karsten

A primitive fungus which is parasitic on many species of Mycena and can be found quite commonly during autumn, covering their hosts with a whisker-like growth.

* **Mycena haematopus** (Fries) Kummer

Cap: 2–4 cm with blood-red flesh, thick at the disc.
Spores: White, elliptical, 7–10 × 5–6 μm. Amyloid.
Stipe: Like *M. sanguinolenta*, it yields a red juice when broken.
Habitat: Usually found growing in clusters on frondose stumps.
Edibility: Not known.

* **Mycena metata** (Fries) Kummer

Cap: 1–2 cm. Conico-campanulate. Flesh-pink or buff. Striate when young.
Gills: Adnate and sub-distant. Whitish to flesh-pink.
Spores: White, 7·5–10 × 4–5 μm. Amyloid.
Stipe: Long, flesh-pink or horn colour. Base white strigose.
Flesh: White, soft and flaccid.
Habitat and season: Usually grows in troops on pine needles or in moss. Very common from September to November.
Edibility: Worthless.

129 Mycena inclinata

130 Mycena oortiana

131 Mycena sepia

132 Mycena sanguinolenta

***Mycena alba** (Bresadola) Kühner
Omphalia alba Bresadola
Marasmiellus albus (Bresadola) Singer

Cap: 0·5–1 cm. Hemispherical, white, striate and minutely pruinose.
Gills: White and broadly adnate.
Spores: White and globose. 6–9 μm. Non-amyloid. Basidia 2- or 4-spored.
Stipe: White and minutely pruinose.
Flesh: White, thin and odourless.
Habitat and season: Grows on mossy bark of living trees. A rare species. From August to November. Not recorded in America.
Edibility: Not known.

133 Mycena pura (Fries) Kummer
Clean mycena

Cap: 2–7 cm. Campanulate, obtusely umbonate, then plane. Very smooth and with a striate margin. Various shades of delicate pink or lilac.
Gills: Adnate connected by veins, subdistant and broad, widest in the middle. Whitish or pink.
Spores: White or cream, ovoid. 6–8 × 3·5–4 μm. Amyloid.
Stipe: Equal or slightly attenuated upwards. Hollow, rigid, polished and tough. Pale pink or lilac with a white hairy base.
Flesh: Thick at umbo, thin at margin. White with a smell and taste of radish.
Habitat and season: Very common locally. On the ground in woods from May to December.
Edibility: Worthless.

* **Mycena pearsoniana** Dennis ex Singer

Cap: 1–2·5 cm. Campanulate then convex, finally plane. Obscurely striate. Rosy or brownish with olivaceous tint.
Gills: Adnate, crowded and narrow. White or olivaceous-white.
Spores: White, pip-shaped and smooth. 5–7 × 3·5–4·5 μm. Non-amyloid.
Stipe: Rigid, straight and more or less equal. Apex pruinose. At first rosy or violaceous-white, then dingy. Base slightly hairy with rooting fibrils.
Flesh: Whitish, mild and slightly rancid.
Habitat and season: Under conifers. Rare. July to November. Not recorded in America.
Edibility: Worthless.

134 Mycena zephirus (Fries) Kummer

Cap: 2–5 cm. Whitish, then reddish or dingy brown. Striate to centre.
Gills: Broad and adnate. White, staining reddish-brown.
Spores: White and cylindrical. 10–12 × 4–5 μm. Amyloid.
Stipe: Equal. Whitish but reddish-brown below. Covered at first with whitish scales. Base roughly hispid.
Flesh: Reddish-brown and thin with no smell.
Habitat and season: Under conifers. Rare. July to November. Not recorded in America.
Edibility: Worthless.

* **Mycena pelianthina** (Fries) Quélet

Cap: 2–4 cm. A violaceous gelatinous species, with distinctly cross-veined gills which have a dark purplish-brown edge.
Spores: Pure white, cylindric-ellipsoid. 5–7 × 2·5–3 μm. Amyloid.
Flesh: Has a faint smell of radish.
Habitat and season: A common species in beech woods. July to November.
Edibility: Not known.

Genus-*Marasmius*

Similar to Collybia but unlike members of that genus in so far as they shrivel in dry weather and revive when moistened. Contains small to medium sized species. The cap is usually tough and pliant. The gills are pliant, usually rather distant and variously attached. The spores are white, smooth and non-amyloid. The stipe is central or wanting, when present then cartilaginous or horny. The odour is often important in identification.

Generally found on woodland debris or on the ground. Some species are edible, probably none is poisonous.

135 Marasmius oreades (Fries) Fries
Fairy-ring champignon or Fairy-ring mushroom

Cap: 2–5 cm. At first convex but finally expanded with a slight umbo. More or less light brown in colour inclined to be darker at centre. When adult the margin is ridged. In moist weather the cap is tacky and the colour brighter, when conditions are dry the cuticle is smooth (kid-like) but wrinkly and the colour appears faded.
Gills: Distant and broad with shorter intermediates. They are rounded towards but free from the stipe and often united by veins. Whitish at first, later pale cap colour.
Spores: White and pip-shaped. 9·5–10·5 × 5·5–6 μm.
Stipe: Cylindrical and slender, long or short, often contorted in dry conditions. Paler than cap, base white and downy.
Flesh: Whitish, thick, tough and elastic. With a pleasant aroma.
Habitat and season: Common in meadows, on golf courses and lawns, especially in sandy areas. From June to November.
Edibility: Excellent when freshly gathered; or it can be dried and stored for future use. The taste is pleasant and aromatic. Care must be taken not to mistake this species with *Clitocybe dealbata* (Fries) Kummer or *C. rivulosa* (Fries) Kummer, both of which are often found growing in close proximity to *M. oreades*. These Clitocybes are very poisonous, sometimes *deadly*, but their gills are crowded and adnato-decurrent, not free.

M. oreades is one of the macro-fungi responsible for causing 'fairy-rings' and can be very troublesome to the groundsman and sportsman alike. Eradication is almost impossible. The best plan is to dig out the areas affected to a depth of at least two feet and refill with sterilized soil, preferably of a different texture and analysis to the original. Another method is to break up or loosen the earth in and around the areas affected (which are often compacted) by using heavy crowbars, then watering in a soil sterilant with an added detergent. There is no guarantee of success with either method.

133 Mycena pura

134 Mycena zephirus

135 Marasmius oreades

136 Marasmius androsaceus

136 Marasmius androsaceus (Fries) Fries
Androsaceus androsaceus (Fries) Rea
Horse-hair fungus

Cap: 0·5–1 cm. Convex but soon expanded. Flattish, wrinkled and striate. Pale rufous, darker at centre.
Gills: Adnate, distant and cap colour.
Spores: White, pip-shaped and smooth. 6·5–9 × 3–4 μm.
Stipe: Horny, smooth and very thin. Shiny-black like horse hair.
Flesh: Virtually none.
Habitat and season: Grows on conifer needles, heather and sticks. Very common from May to November.
Edibility: Worthless.

⋆ **Marasmius undatus** (Berkeley) Fries

Grows on bracken stems.

⋆ **Marasmius epiphylloides** (Rea) Saccardo & Trotter

Grows on ivy leaves.

⋆ **Marasmius hudsonii** (Fries) Fries

Grows on holly leaves.

⋆ **Micromphale perforans** (Fries) S. F. Gray
Marasmius perforans (Fries) Fries

Grows on pine needles.

137 Marasmius scorodonius (Fries) Fries

Cap: 1–2 cm. Convex but soon plane and slightly gibbous. Always dry. Even and rufous at first but soon becoming wrinkled, paler and whitish.
Gills: Distant, narrow and adnate but often parting company with the stipe. Connected by veins and finally wrinkled. Whitish.
Spores: White and elliptical. 7–9 × 3·5–4·5 μm.
Stipe: Thin, tall, equal and rooting. Glabrous and shining, horny and hollow. Rufous in colour.
Flesh: Whitish, thin and tough. With a strong odour of garlic.
Habitat and season: Common amongst plant debris, twigs etc., especially beech. From August to November.
Edibility: Unsubstantial but can be used either fresh or dried as a substitute for garlic.

* **Marasmius alliaceus** (Fries) Fries

Similar to *M. scorodonius* but with minutely velvety black stipe. The ovate spores are 9–11·5 × 6–7 μm. It also has a strong smell of garlic.

138 Marasmius cohaerens (Fries) Cooke & Quélet
Mycena balanina (Berkeley) Karsten

Cap: 2–3 cm. Campanulato-convex. Smooth but velvety in appearance and somewhat fleshy. Tawny, later paler.
Gills: White. Rather distant, rounded behind and often connected by veins.
Spores: White, pip-shaped, curved at base. 9–10 × 5 μm.
Stipe: Equal, rather horny and shining. Dark brown, paling to whitish at apex. Covered with long fine hairs at base.
Flesh: White in cap, brownish in stipe. Thick at disc, thin at margin.
Habitat and season: Common and gregarious in mixed woodlands. August to November.
Edibility: Worthless.

* **Marasmius ramealis** (Fries) Fries
Marasmiellus ramealis (Fries) Singer

Cap: 0·5–1 cm. Convex then plane. Slightly wrinkled. Pinkish-white.
Gills: Sub-distant, narrow and connected behind. Whitish.
Spores: White and elliptic. 8–10 × 2·5–3·5 μm.
Stipe: Equal and usually curved. Whitish with rufescent base.
Flesh: Thin and whitish.
Habitat and season: Densely gregarious on dead twigs etc. Common from June to November.
Edibility: Worthless.

* **Marasmius rotula** (Fries) Fries
Androsaceus rotula (Fries) Patouillard

Cap: 0·5–1·5 cm. Convex to plane. Margin scalloped and with umbrella-like radiating folds. Whitish.
Gills: Very distant, equal and connected to a collar which is free from the stipe. White.
Spores: White and pip-shaped. 7–10 × 3·5–5 μm.
Stipe: Very slender, horny and shining. Deep brown or blackish.
Flesh: White in cap.
Habitat and season: Usually gregarious on dead twigs etc. Common throughout the year in suitable weather.
Edibility: Worthless.

Genus-*Amanita*

Large and terrestrial fungi which, in the early stages, are completely enclosed within a universal veil. With growth, this veil is ruptured leaving a portion at the base of the stipe (the volva). The other portion usually remains on the cap forming scales or patches (the velar remains). The gills are almost always crowded and free, with whitish spores. The stipe is central and furnished with a ring which is a remnant of the partial veil. The flesh of the stipe is usually distinct from that of the cap. The volva, the ring and the velar remains on the cap can be evanescent. Also there are a very few species in which the median ring is absent from the stipe; these are sometimes referred to under a separate genus—*Amanitopsis*.

The genus includes deadly and dangerously poisonous species, as well as some of the finest esculents. It is, therefore, important that the mycophagist should become well-acquainted with the characteristics of this group.

139 Amanita citrina Schaeff. ex S. F. Gray
Amanita mappa (Lasch.) Quélet
False death cap

Cap: 5–9 cm. Hemispherical then expanded; fleshy. Usually bearing patches of velar remains. The general colour is pale lemon-yellow.
Gills: Adnexed, narrow and crowded. White, sometimes with yellowish edge.
Spores: White, smooth and egg-shaped. 8–10 × 7–8 μm. Amyloid.
Stipe: White or yellowish-white. Robust, tapering towards the apex and with a distinctly bulbous base.
Flesh: White and ample. Smells of raw potato. When treated with Nitric acid, immediately becomes intense blue.
Ring: White often with yellow tints. Superior and lax from upper part of stipe.
Volva: Upper limb short, forming a gutter around top of the large bulbous base.
Habitat and season: Common on the ground in all types of woodland. July to November.
Edibility: Worthless; has a disagreeable taste but not poisonous.

137 Marasmius scorodonius

138 Marasmius cohaerens

139 Amanita citrina

140 Amanita muscaria

There is also a wholly white variety *alba* which is common under beech trees.

140 Amanita muscaria (Fries) Hooker POISONOUS
Fly agaric

Cap: 6–20 cm across hemispherical at first, later expanding. Scarlet, sometimes orange-red and usually covered with numerous white scales or warts left by the torn veil. The colour fades with age or washes out to a pale orange colour. Height up to 25 cm.
Gills: White, crowded, broad and free from the stipe. Interspaced with intermediate gills.
Spores: White, ovate. 9–12 × 6·5–8 μm.
Stipe: White with bulbous base and volva. Often hollowing with age. There is a large white, membranous hanging ring which may have vertical striae.
Flesh: White but red underneath cap cuticle. Mild smell.
Habitat and season: A woodland species, particularly birch and conifer. Common from August to November.
Edibility: *Not to be eaten.* Though poisonous, it is doubtful if it would cause the death of a healthy person.

It was believed, in medieval times, that if broken into pieces and placed in milk it would repel or kill flies, hence its common name *Fly agaric*. This fungus has been used as an intoxicant by a few primitive tribes of Eastern Siberia.

141 Amanita rubescens (Fries) S. F. Gray
The Blusher

Cap: 6–12 cm. Hemispherical, then convex and finally flat. Reddish fawn or brown and covered with dirty-white warty patches (velar remains) which are soon washed away in wet weather. (Our illustration is of a specimen in prime condition.) The margin may show striation when old.
Gills: White but spotted reddish with age. Crowded and thin, narrowing towards the stipe and attached to it by a small tooth.
Spores: White and elliptical. 9–10 × 5–6 μm.
Stipe: White with a pinkish flush especially towards the bulbous base. Sturdy and tapering upwards. Striate above the ring. Bruises reddish.
Flesh: White but soon becomes reddish when infested with insect larvae to which it seems particularly prone.
Ring: White. Superior, large and lax.
Volva: Inferior, amounting only to scales around the base of stipe.
Habitat and season: Very common in frondose and coniferous woods. From July to November.
Edibility: Excellent when cooked but can cause indigestion and sickness if eaten raw. Identification *must be certain* owing to the superficial resemblance to *A. excelsa* and *A. pantherina* the latter being *very poisonous.*

142 Amanita excelsa (Fries) Kummer
Amanita spissa (Fries) Kummer

Cap: 6–12 cm. Convex at first, later expanded. Of variable ground colour, brown, fuliginous or grey. Compact and smooth, clothed with small cinereous angular adnate warts.
Gills: Free but reaching stipe. Broad, crowded and shining white.
Spores: White and ovate. 8–11 × 5·5–8 μm. Amyloid.
Stipe: Of variable shapes. Having a bulbous base and more or less white in colour. Striate above the ring which is set high, superior and lax. Volva is indistinct.
Flesh: Firm, white and unchanging.
Habitat and season: Found fairly commonly in all types of woodland from June to October.
Edibility: Said to be harmless but is easily mistaken for the very poisonous *A. pantherina* which it closely resembles. Because of this we consider it should not be eaten.

* **Amanita pantherina** (Fries) Secretan — POISONOUS
Panther cap or False blusher

Cap: 5–8 cm. Convex, then expanded. Brown but often tinged olive. Viscid or dry and having pure white warts. Margin striate.
Gills: White, crowded and free.
Spores: White and ovate. 9–12 × 6·5–8 μm. Non-amyloid.
Stipe: White and variable in length, tapering upwards. Bulbous and soon becoming hollow. The ring is usually set low down by comparison with *A. excelsa*, often residual volva remnants between bulb and ring.
Flesh: White and unchanging. If treated with a 10% solution of potassium hydroxide turns orange-yellow in cap.
Habitat and season: Occasionally found in frondose woods especially beech. August to October.
Edibility: *Poisonous.* Sometimes *deadly.*

143 Amanita phalloides (Fries) Secretan — DEADLY POISONOUS
Death cap

Cap: 7–12 cm. Hemispherical at first then convex and finally flattened. Slightly viscid when wet. Usually without velar remains. Colour variable: yellowish-olive, greenish, greyish-white etc. Streaked with darker radiating fibrils.
Gills: Free, crowded and white. When treated with concentrated sulphuric acid they turn pinkish-lilac.
Spores: White, smooth and elliptical with a large central oil drop. 9–11 × 7–9 μm. Amyloid.
Stipe: Bulbous at base, tapering somewhat upwards. Stuffed, then hollow. Smooth or floccose. White but sometimes with a greenish tinge.
Ring: Superior, lax and white. Striate above.
Volva: Bag-like with irregular lobes. White or white with yellowish-greenish tinges externally.
Flesh: White but yellowish-greenish under cap cuticle which can be readily peeled. Odour is mild when young, becoming strong and fetid with age.
Habitat and season: Fairly common in frondose woods and adjoining fields. July to November.
Edibility: *Deadly poisonous* even in very small quantities. It is responsible for most fatalities caused by fungi. The symptoms do not become evident until digested or even up to 24 hours after ingestion. By this time emetics are useless.

It is difficult to understand why the mycophagist should make the dreadful mistake of taking this fungus for culinary purposes as it bears no resemblance to the edible mushrooms of our countries and its characteristics are specific.

* **Amanita virosa** Secretan — DEADLY POISONOUS
Destroying angel

Cap: 5–9 cm. White. Convex and slightly umbonate. Viscid and shining smooth. Turns chrome yellow when treated with Potassium hydroxide (KOH).
Gills: Crowded and free. White.
Spores: White and globose. 8–10 μm. Amyloid.
Stipe: Slender and floccose.
Ring: Superior and lax, also frequently oblique to the stalk.

141 Amanita rubescens

142 Amanita excelsa

143 Amanita phalloides

144 Amanita inaurata

Volva: Thick and envelopes the basal bulb but is detached from it at the top.
Flesh: Thin, soft and white.
Habitat and season: A species of frondose woods, particularly northern, from August to October.
Edibility: Like *A. phalloides*, it is *deadly poisonous*.

144 Amanita inaurata (Secretan)
Amanita strangulata (Secretan) Boudier
Amanitopsis strangulata (Fries) Roze apud Karsten
Ringed-foot

Cap: 8–12 cm. Grey-brown. Similar to *A. vaginata* but larger, much more robust and squat. Cap often carries large patches of velar remains.
Gills: Close, free and remote from stipe. Creamy or greyish-white.
Spores: White and globose. 10·5–13 μm. Non-amyloid.
Stipe: White or very light brown. No ring. Striped by the brown or greyish flocci which is distributed along its entirety.
Flesh: White. Without odour.
Volva: White at first, grey later. Thick but fragile and soon fragmented.
Habitat and season: Locally common in mixed woodlands and heaths especially on substrates with high alkalinity. August to November.
Edibility: Good, but *must be correctly identified first*.

145 Amanita fulva Secretan
Amanitopsis fulva (Secretan) W. G. Smith
Tawny grisette

Cap: 4–7 cm. Semi-ovate at first soon expanding, often slightly depressed with umbo. Splitting at margin which is noticeably striate. Orange-brown becoming browner, occasionally with white velar remains.
Gills: Pure white, free, rather crowded but neatly spaced.
Spores: White, globose, 9–11(12) μm. Non-amyloid.
Stipe: Long and slender, tapering upwards. White, thinly flocculose at first. Tinged tawny especially near base. *No ring.*
Volva: White and bag-like with a ragged top-edge which is tinged orange-brown.
Flesh: White, tender and fragile.
Habitat and season: Very common in frondose and mixed woodlands from May to November especially during the months of autumn.
Edibility: Good, having a mild taste.

* **Amanita vaginata** (Fries) Vittadini
Common grisette

Like *A. fulva* but cap is light or dark grey. Stipe whitish-grey, and volva is tinged with grey at top edge. *No ring.*

146 Amanita aspera (Fries) S. F. Gray POISONOUS

Cap: 5–8 cm. Convex then plane, yellowish to dusky-olive. Covered with minute, closely crowded, sulphur-coloured warts. Margin thin, not striate.
Gills: White and free. Broad in front, rounded behind and ventricose. Yellowing with age.
Spores: White. 8 × 6 μm.
Stipe: Stuffed. Short at first then elongating and attenuating upwards from a wrinkled bulb covered with sulphur-yellow scales.
Ring: Superior and entire, margin has sulphur-yellow flocci and is striate above.
Volva: Free, margin obsolete.
Flesh: Thick and white but inclining to cap colour under cuticle. Often attacked by insect larvae which causes a reddening of the flesh.
Habitat and season: Occasional in frondose woods, especially beech. From August to November.
Edibility: *Poisonous.*

147 Amanita porphyria (Fries) Secretan
Amanita recutita (Fries) Gillet

Cap: 5–9 cm. Convex, smooth and greyish-brown with a purple tinge.
Gills: Creamy-white, crowded and free.
Spores: White and sub-globose. 7·5–9·5 × 6·5–7·5 μm. Amyloid.
Stipe: White, longish and cylindrical with a large bulb and a distinct volva. The ring which is set high is tinged greyish-brown on the underside.
Flesh: White and unchanging.
Habitat and season: Occasionally found in mixed woodlands, more especially coniferous. September to October.
Edibility: Whilst not poisonous, its unpleasant taste renders it inedible.

148 Amanita gemmata (Fries) Gillet
Amanita junquillea Quélet

Cap: 5–11 cm. Hemispheric, campanulate then plane and maybe centrally depressed. Slightly viscid. Margin thin and striate. Yellow, ochraceous or whitish, more intense at disc. With or without whitish floccose warts.
Gills: White and crowded. Just free. Broad in the middle.
Spores: White. 7·5–11 × 6–9 μm. Non-amyloid.
Stipe: White and fibrillose or concolorous with the cap. Slender or stout and with a bulbous base.
Ring: White and lax, eroding with age.
Volva: White. Attached to basal bulb forming a collar round it.
Flesh: White and tender. Yellowish under cap cuticle. Has a mild flavour.
Habitat and season: On the ground in woodlands. From April to November.
Edibility: Pleasant tasting but not recommended as some people find it indigestible.

* **Amanita verna** (Fries) Vittadini DEADLY POISONOUS
Amanita phalloides var. *verna* (Fries) Rea
Spring amanita

Cap: 5–10 cm. Oval then expanded, finally slightly depressed. Glabrous, viscid in moist weather, glossy when dry. Margin thin and only striate with age. White becoming pale ochraceous at the disc later.
Gills: Crowded and with intermediates, free. White.
Spores: White and ellipsoid. 8–11 × 7–9 μm.
Stipe: Tall and slender, thickening gradually downwards to a bulbous base. Stuffed then hollow. White and flocculose.
Ring: Superior, membranous and striate. White.
Volva: Firm and enveloping the bulb with free-lobed margin. White and membranous.
Flesh: Tender and white. Odour insignificant when young but becoming unpleasant of old potatoes with age. Taste slightly acrid but *we do not advise tasting this!*
Habitat and season: Uncommon in mixed woodlands late summer to mid autumn. The common name is very misleading. Vittadini misinterpreted this fungus. It is *not* a spring species.
Edibility: *Deadly poisonous.*

* **Amanita solitaria** (Fries) Secretan
Amanita strobiliformis (Vittadini) Quélet

Cap: 6–20 cm. Sub-globose at first then convex and finally

145 Amanita fulva

146 Amanita aspera

147 Amanita porphyria

148 Amanita gemmata

expanded. White to light grey or yellowish-grey. Smooth and covered with white floccose warts which soon become darker and hardened, they are easily removed by wind and rain. When young the margin is often festooned with velar remains.
Gills: Crowded and broad, adnexed or free, narrow behind. White.
Spores: White and ellipsoid. 10–12·5 × 8–10 μm. Amyloid.
Stipe: Thick and solid with a rooting bulbous base. Cap-colour with mealy white scales which are easily rubbed off.
Ring: Superior, lax and tomentose. White and delicate, eroding quickly.
Volva: Indistinct.
Flesh: White, thick and tender. Has a pleasant taste and mild odour.
Habitat and season: Grows singly or in groups under frondose trees and in the adjacent fields, especially on base rich soil. A rare species which occurs from July to October.
Edibility: Good when young, but cap cuticle should be removed before cooking.

* **Amanita caesarea** (Fries) Quélet
Caesar's mushroom

Cap: 8–18 cm. Ovoid, then convex, finally plane. Only rarely with white velar remains. Margin striate. Usually orange, reddish-orange or yellowish-orange (not scarlet). Viscid in humid conditions. The cuticle is easily peeled.
Gills: Crowded and free. Yellow (not white).
Spores: White to tinged yellowish, ellipsoid, 10–14 ×

6–11 μm. Non-amyloid.
Stipe: Yellow, cylindrical and thick. Has a bulbous base which is enveloped by the volva.
Ring: Yellow. Superior, lax and usually striate.
Volva: White. Large, lobate and free from stipe at the top edge.
Flesh: Firm and white. Yellowish under cap cuticle. With a pleasant taste and odour.
Habitat and season: Under frondose trees, especially oak, chestnut and walnut; rarely in coniferous woods. From July to November in parts of Europe and the Americas. Not as yet found in Britain.
Edibility: Excellent. Said to be the best fungus—raw, cooked or preserved in oil.

Great care should be taken when collecting this esculent, since it can be mistaken for the poisonous *Amanita muscaria*.

* **Amanita brunnescens** Atkinson DEADLY POISONOUS
Browning amanita

Cap: 3–15 cm. Hemispherical at first, then convex, finally plane. Slightly viscid when wet. Margin covered with white warts and patches, also faintly striate. Dark brown or olivaceous-brown.
Gills: Crowded, free and broad. White, somewhat floccose or crenulate at the edge.
Spores: White, round and thin walled. 7–9·5 μm.
Stipe: Enlarged downward, with a large marginate bulb which is conspicuously split or cleft longitudinally one or more times. Fibrillose and slightly striate above. White but often becoming brownish at apex.
Ring: Superior, membranous and pendant. White.
Volva: At first adhering in irregular patches to the basal bulb, often evanescent with age.
Flesh: Soft, fragile, thin and white.
Habitat and season: Common in frondose and mixed woods in parts of N. America. Not recorded in Britain.
Edibility: *Deadly Poisonous.*

Genus-Volvariella

Medium-sized species with cap fleshy and regular. The gills are free and becoming deep pink. The spores are smooth, broadly oval and deep pink. Stipe is central with a distinct volva but no ring. The flesh of the cap and stipe not continuous.

* **Volvariella speciosa** (Fries) Singer
Volvariella gloiocephala (Fries) Gillet
Handsome volvaria

Cap: 5–10 cm. Campanulate then expanded and rather gibbous. Smooth and glabrous, viscid in damp conditions only. White to ashy-grey.
Gills: Crowded, broad and free. White then pink, finally brownish-pink.
Spores: Pink. 13–18 × 8–10 μm.
Stipe: Long and firm, whitish and smooth. Attenuated upwards.
Volva: Membranous, bulbous and free. Externally tomentose. Whitish.
Flesh: Tender and white with a mild taste and odour.
Habitat and season: On organically rich ground, dunghills and compost heaps etc. Occasional from July to October.
Edibility: Good but identification must be positive.

* **Volvariella bombycina** (Fries) Singer
Silky volvaria

Cap: 5–20 cm. Ovoid then campanulate and finally expanded. White and covered with very fine silky fibrils or squamules of white or yellowish-white.
Gills: Crowded and free, ventricose with eroded edge. White at first then pink and finally ochraceous-pink.
Spores: Pink. 7–9 × 5–6 μm.
Stipe: Firm and solid, often curved. Attenuated gradually upwards from a bulbous base.
Volva: Tall and very large. Lax and membranous with an uneven lobate margin. Persistent and somewhat viscid. White then dingy.
Flesh: Thin, tender and white. Has an odour of wood and a pleasant taste.
Habitat and season: Usually solitary but sometimes caespitose on decomposing stumps of frondose trees, especially elm; also on sawdust. A rare and handsome species found from June to October.
Edibility: Very good.

* **Volvariella surrecta** (Knapp) Singer
Volvaria surrecta (Knapp) Ramsbottom
Volvaria loveiana (Berkeley) Gillet

Cap: 2–5 cm. Silky white or slightly cinereous.
Gills: Pink.
Spores: Pink and smooth. 5–6 × 3–4 μm.
Stipe: Attenuated upwards and often curved from a slightly bulbous base. Fibrillose, solid and white.
Habitat and season: Parasitic on the caps of decaying and generally distorted specimens of *Clitocybe nebularis* and *C. clavipes*. Appearing at first as smooth round swellings 1–2·5 cm across. *Rare.*
Edibility: Not known. Berkeley stated that it has a taste of the field mushroom.

* **Volvariella parvula** (Weinm.) P. D. Orton POISONOUS
Volvariella pusilla (Fries) Singer

Cap: 1–2·5 cm. Conical then campanulate and finally plane with an umbo. Silky white.
Gills: Crowded and free. Pale pink.
Spores: Pink, 5·5–6·5 × 4·5 μm.

Stipe: Equal, silky and white with a hairy base. Internally pithy then hollow.
Volva: Minute, membranous and lax. Often 3- to 4-lobed.
Flesh: White and soft.
Habitat and season: Gregarious in woods, gardens and greenhouses etc. Occasional from June to October.
Edibility: *Poisonous.*

* **Volvariella media** (Fries) Singer POISONOUS

Similar to, but usually larger than, *V. parvula*.
Cap: 2·5–5 cm. viscid.
Gills: Crowded and free, slightly ventricose. Pale pink.
Spores: 5 × 3·5 μm.
Stipe: Solid, not pithy nor hollow.
Habitat and season: Woods and pastures. June to October.
Edibility: *Poisonous.*

Genus-*Pluteus*

Large to small species, centrally stipitate. The flesh of the cap is distinct from that of the stipe. Ring and volva absent. The gills are free and usually salmon-pink at maturity. The spores are smooth, oval and deep salmon or rosy ochre. On wood or terrestrial.

* **Pluteus cervinus** (Fries) Kummer
Pluteus curtisii (Berkeley & Broome) Saccardo
Fawn-coloured pluteus

Cap: 5–10 cm but can be more than double this size; it is easily detached from the stipe. Campanulate then expanded, often cracking. Innately radially fibrillose. Viscid in humid weather. It occurs in varying shades of brown and is usually darker at the umbo.
Gills: Crowded and free, thin and broad, rounded towards the stipe. White at first then pink, finally brownish.
Spores: Rosy-ochre, broadly elliptical and smooth. 7–8 × 5–6 μm.
Stipe: Fairly long, solid and equal or sometimes swollen at the base. White but appearing greyish due to dark fibrils.
Flesh: White. Thick at cap disc but thin elsewhere. Has a mild odour of turnips.
Habitat and season: Grows singly or is gregarious on rotten stumps and fallen trunks; may also be found on sawdust heaps. Common from May to November but occurs during mild weather at any time of the year.
Edibility: Of poor quality.

* **Pluteus petasatus** (Fries) Gillet

Cap: 5–15 cm. Convex then expanded. Fleshy and whitish becoming brownish, especially with age. Viscid.
Gills: Very crowded, free. White then deep salmon.
Spores: Rosy ochre and ovoid. 7–9 × 4–5 μm.
Stipe: Stout and whitish, narrowing downwards. Sulcate at base.
Flesh: White and ample.
Habitat and season: On sawdust. Rare.
Edibility: Not known.

Genus-*Lepiota*

Large to small fungi usually found on the ground or on manure and compost heaps. Universal veil present in the early stages. Cap is scaly or mealy, rarely smooth and viscid. The gills are white or whitish (when young) and free. Exceptions are: in the closely related Melanophyllum, the gills may also be green (*M. eyrei*) or even reddish (*M. echinatum*). They are often very remote from the stipe and attached to a collar.

Spores are typically whitish (dextrinoid), except: *M. echinata* whose spores are red; *L. morgani*, *L. molybdites* and *M. eyrei* all of which have greenish spores.

The stipe is centrally placed with either a single or double ring, which is persistent and moveable or evanescent. The flesh is usually distinct from that of the cap but there are exceptions.

Volva absent, and this is an important guide to identification.

The majority of species are edible (excellent or good) but a very few are considered poisonous or suspect.

149 **Lepiota sistrata** (Fries) Quélet
Lepiota seminuda (Lasch.) Kummer
Cystoderma seminudum (Lasch.) Singer

Cap: 2–3·5 cm. Campanulate then expanded and often obtusely umbonate. Mealy with shining particles. Whitish, tinged yellow or pinkish and with a darker disc. When young the margin is festooned with velar remains.
Gills: Crowded, ascending and almost free. Clear white.
Spores: White and elliptical. 3–4 × 2–2·5 μm.
Stipe: Equal and flexuose. Loosely stuffed with fibrils. Silky-white or flesh colour, reddish below.
Ring: Fibrillose, quickly evanescent.
Flesh: Thin and white.
Habitat and season: Frequent especially on sandy soils. From August to October.
Edibility: Yes, but identification must be certain.

150 **Lepiota excoriata** (Fries) Kummer

Cap: 6–10 cm. Globose, finally expanded with a slight umbo. Adult cuticle often drawn up and falling short of margin. Pale fawn, but with an expansion of the cap, the thin cuticle breaks up into scales, between which the background colour is lighter and the substance silky. The umbo remains dark.
Gills: White, soft and crowded. Free from the stipe by a collar.
Spores: White and elliptic with germ pore. 12–16 × 8–10 μm.
Stipe: About the same length as the cap diameter and more or less equal. Has a slightly bulbous rooting base. Whitish, minutely fibrillose and stuffed with a cottony substance, then hollow.
Ring: Deflexed, moveable and white.
Flesh: Thick and white, soft in cap. Unchanging.
Habitat and season: Common in permanent pastures with light soil. Usually in fair numbers. August to October.
Edibility: Good.

151 **Lepiota clypeolaria** (Fries) Kummer
Shield lepiota

Cap: 3–8 cm. Campanulate then plane. Margin often with velar remains. Background white but covered with a yellow or brown felt-like layer which breaks up into patches.
Gills: Crowded and just free. Soft, white or yellowish.
Spores: White and fusiform with one or more oil drops. 13–19 × 5–6 μm.
Stipe: Slender, soft and fragile. Equal or slightly thickened at base. Colour as cap. Felty below the ring when young, but becoming almost naked. Hollow with age.
Ring: Soft and woolly; soon eroding.
Flesh: White, soft and thick.
Habitat and season: On the ground in woods, gardens and hothouses etc. From September to November. A very variable species depending on its environment.
Edibility: Opinions vary as to the edibility of this species. In North America it is said to be *poisonous*. Our advice is do not take the risk.

* **Lepiota fuscovinacea** F. H. Møller & J. Lange

Cap: 3–4 cm. Umbonato-convex. Vinous-red to brown. Felty with darker scales.
Gills: Free and fairly crowded. Whitish becoming dingy.
Spores: White and ovoid. 4–5·5 × 2–3 μm.
Stipe: Stout, fibrillar and velvety. Dark brown below, paler at apex.
Ring: Membranous and dirty white.
Flesh: Fairly thick. Dirty white to brown.
Habitat and season: In damp places under trees, in ditches etc. Very sombre in appearance. A rare species. From September to October.
Edibility: No.

152 **Lepiota rhacodes** (Vittadini) Quélet
Shaggy parasol

Cap: 8–15 cm when expanded. Ovate at first but broadly umbonate when mature. The ground colour is white covered with large stiff, ragged, brownish scales concentrically arranged in rows and raised at their tips. The umbo is dark shining brown and the margin fringed.
Gills: Crowded, free by a collar. Bruising reddish at the touch.
Spores: White, elliptical and smooth, with a germ-pore. 9–12 × 6–7 μm.
Stipe: Dirty white, smooth and stout with a large basal bulb. May be longish or short but tapering upwards. Bruising reddish. Internally it is pithy or hollow.

149 Lepiota sistrata

150 Lepiota excoriata

151 Lepiota clypeolaria

152 Lepiota rhacodes

Ring: Double and free. Greyish-white then reddish.
Flesh: White and tender then brownish. It turns reddish when cut, especially in stipe.
Habitat and season: Occasional in woods, gardens and compost heaps. Usually gregarious. Found July to November.
Edibility: Very good but the stipe should be discarded.

* **Lepiota badhamii** (Berkeley & Broome) Quélet

Cap: 3–10 cm. It is less scaly than *L. rhacodes* but shares the same characteristic of bruising reddish to the touch. Has a spindle-shaped stipe.
Edibility: Yes.

* **Lepiota leucothites** (Vittadini) P. D. Orton
Lepiota naucina var. *leucothites* (Vittadini) Saccardo
Nutshell lepiota

Cap: 6–10 cm. Wholly white but occasionally with a slight pink or yellowish tinge. At first the cuticle is smooth and silky, later becoming granular.
Gills: White, crowded and free by a collar. Finally flesh coloured.
Spores: White, with large germ pore. 8–9 × 5 μm.
Stipe: Equal and smooth.
Ring: High up the stipe, narrow, free and superficial.
Flesh: White, soft and ample.
Habitat and season: Found in fields and gardens during summer and autumn. Occasional.
Edibility: Yes.

153 Lepiota cristata (Fries) Kummer POISONOUS
Crested lepiota

Cap: 2–4 cm. Bell-shaped then plane or with margin slightly upturned but centre remains umbonate. White and silky with great numbers of minute, crowded, reddish-brown scales which peter out towards the margin.
Gills: White, very crowded and free.
Spores: White and oblong but spurred on one side near the base. 6–8 × 3–4 μm,
Stipe: Long or short but usually curved. More or less equal, thickened at base. White or tinged brownish, silky fibrillose and fragile.
Ring: White and superficial, situated about half way up the stipe.
Flesh: Whitish and thin, with an unpleasant taste and odour.
Habitat and season: In groups in lawns, pastures and woods, forming sclerotia in the substrate. Common from August to November.
Edibility: *Poisonous.*

154 Melanophyllum eyrei (Massee) Singer
Lepiota eyrei (Massee) J. Lange
Green-spored lepiota

Cap: 1–4 cm. Umbonato-convex, becoming plane with slight umbo. Mealy at first with velar remains on margin. White to greyish-brown, darker at disc, finally completely dirty white or grey.
Gills: Crowded and free. Pale bluish-green becoming verdigris green.
Spores: Greenish in the mass and oval. 3·5–5 × 2–2·5 μm.
Stipe: Thin, fragile, mealy granular and hollow. Grey to brownish, darker towards the base.
Ring: Very inferior and evanescent.
Flesh: Thin and whitish.
Habitat and season: On base rich soil, often in moss under trees. From September to October.
Edibility: Unknown. Suspicious.

Genus-*Cystoderma*

Small to medium-sized species with cap and lower portion of stipe covered with granules.
Formerly referred to under Lepiota.

155 Cystoderma amianthinum (Fries) Fayod
Lepiota amianthina ([Scop.] Fries) Karsten

Cap: 2–5 cm. Convex, then plane and somewhat umbonate. Granulosely scurfy, margin denticulate. Yellow to ochre. Cuticle becomes rusty-brown when treated with a 10% solution of KOH Potassium hydroxide.
Gills: Adnate and fairly crowded. White, then whitish-yellow.
Spores: White and elliptical. 5–7 × 3–4 μm.
Stipe: Slender and equal, covered with yellow squamules up to the ring. Yellow or brownish, and smooth above ring.
Ring: Pale and superficial.
Flesh: Yellow, thin and fragile.
Habitat and season: Fairly common in the mossy areas of pine woods and heaths. From August to November.
Edibility: Yes, but not recommended.

There is also a pure white variety.

* **Cystoderma granulosum** (Fries) Fayod
Lepiota granulosa (Batsch ex Fries) S. F. Gray

Similar to *C. amianthinum* but usually larger.

Cap: Rufous brown. Cuticle becomes rusty-brown when treated with 10% KOH, Potassium hydroxide.
Gills: Slightly adnexed.
Spores: Oval. 3·5–5 × 2·5–3 μm. Non-amyloid.
Stipe: Stout.
Ring: Inconspicuous.
Habitat and season: Under conifers, August to October.
Edibility: Worthless.

* **Cystoderma cinnabarinum** (Secretan) Fayod
Lepiota cinnabarina (Alb. and Schw. ex Sec.) Karsten

Cap: 5–8 cm. Persistently brick-red or cinnabar, and granulosely scurfy.
Gills: Free and white.
Spores: 6–7 × 5 μm.
Stipe: With red scales up to ring-zone but paler and smoother above.
Ring: Inconspicuous.
Habitat and season: Heaths and woods in autumn.
Edibility: Worthless.

Genus-*Agaricus*

The cap is fleshy and regular, white and smooth or brownish and scaly. The gills are crowded and free. Whitish, grey or deep pink at first but finally becoming dark brown. The spores are dark brown and smooth, the stipe is central and fleshy fibrous with a ring which often erodes. All are terrestrial.

156 Agaricus langei (F. H. Møller) F. H. Møller

Cap: 7–10 cm. Convex and flattish at disc, then expanded. With thickly set adpressed reddish-brown scales.
Gills: Crowded and free. Deep pink at first but finally dark brown.
Spores: Purple-brown. 7–8(9) × 4·5–5 μm.
Stipe: Stout and whitish but soon dingy.
Ring: White and large at first, becoming dingy and eroding.
Flesh: White but turns bright red when cut. Has a pleasant taste and odour.

153 Lepiota cristata

154 Melanophyllum eyrei

155 Cystoderma amianthinum

156 Agaricus langei

Habitat and season: Gregarious under frondose trees. Occasional from September to November.
Edibility: Very Good.

* **Agaricus arvensis** J. Schaeffer ex Secretan
Horse mushroom

Cap: Up to 20 cm across when expanded but at first semi-ovate and flattened at umbo. Creamy-white bruising yellowish.
Gills: Crowded, free and rather narrow. Greyish then pale brown, finally dark brown.
Spores: Dark purple-brown, elliptical and smooth. 6·5–8 × 4·5–5·5 μm.
Stipe: Tall and thick, equal or slightly swollen at base.
Ring: Superior, appearing double.
Flesh: Thick and firm in the cap. Odour of aniseed.
Habitat and season: In pastures and often on roadside verges. July to August.
Edibility: Excellent.

* **Agaricus xanthodermus** Genevier
Psalliota xanthoderma (Genevier) Richen & Roze

* **Agaricus placomyces** Peck
A. meleagris Schaeff.
Yellow-staining mushroom

These are two species with the outward characteristics of *A. arvensis* but often cause sickness when eaten. If the extreme base of the stipe is cut, in either species, there is an immediate change of colour to bright yellow. Spores are much smaller.

157 Agaricus campestris Fries
Psalliota campestris (Fries) Quélet
Field mushroom

Cap: 4–10 cm. White and dry. Hemispherical then convex, finally expanded. Smooth to slightly fibrillose or even scaly. Velar remains are often found adhering to the margin.
Gills: Crowded and free. At first white to pink, then milk chocolate and finally very dark brown.
Spores: Dark brown. 7–8 × 4·5–5·4 μm.
Stipe: Cylindrical or tapering downwards. Solid and white with a ring that soon erodes or falls away.
Flesh: Thick and white but faint reddish tinge above gills. A faint reddish tinge also occurs when the flesh is infested with insect larvae.
Habitat and season: Common in pastures, meadows and lawns etc. Usually growing in groups or circles from July to November.
Edibility: Excellent when cooked but may be eaten raw.

158 Agaricus silvicola (Vittadini) Peck
Psalliota silvicola (Vittadini) Ricken
Pratella flavescens Gillet
Wood mushroom

Cap: 6–12 cm. Umbonate then expanded or plane. White, shining and smooth. Bruising yellow or becoming yellow with age.
Gills: Crowded, distant and rather broad. Greyish-white and tinged pink, becoming brown then dark purple-brown.
Spores: Purple-brown, smooth and elliptical with central oil drop. 5–6 × 3·5–4 μm.
Stipe: Usually long with swollen or bulbous base. Smooth and white, yelllowing and becoming hollow with age.
Ring: Ample and lax, high on stipe, sometimes double. White above, drab below.
Flesh: White to cream and rather unsubstantial, purplish adjacent to gills. Has odour of anise.
Habitat and season: On the ground in woods and in fields adjacent to woods. Fairly common from August to November.
Edibility: Good.

* **Agaricus augustus** Fries
Psalliota augusta (Fries) Quélet
The Prince

Cap: 10–20 cm. Almost globose with flattened apex then convex to plane, very obtuse. Minutely fibrillose, squamulose towards the margin. Ground colour yellowish-brown, squamules darker.
Gills: Crowded, free and distant from stipe, narrow. Whitish then brown, never pink.
Spores: Purple-brown in the mass and elliptical. 7–10 × 4·5–5·5 μm.
Stipe: Tall and slightly attenuated upwards. Whitish and scaly below ring when young, becoming smooth and yellowish. Bruising yellowish.
Ring: Superior and membranous, white and lax.
Flesh: White and ample, yellowish with age. Odour of aniseed.
Habitat and season: Occasional under coniferous and frondose trees. Fries states that it often grows on ant-hills. Found August to November.
Edibility: Very good.

Genus-Coprinus

Large to small fragile species, centrally stipitate with a fleshy or membranous cap which at first is cylindrical or ovate, and scaly or mealy. The margin is adpressed to the stipe which in some species has a ring. The gills are adnate, free or attached to a collar. They are very thin, parallel sided and close together when young and in most cases quickly deliquescing (auto-digesting) into a thick black fluid. The auto-digestion commences at the cap margin and continues inwards or upwards as the case may be, often the entire cap is so affected; it coincides with the ripening of the spores and renders assistance in their discharge.

A few species in the genus do not auto-digest but in all other respects they show the same characteristics.

The spores are black, very dark or violaceous-brown in the mass. They are smooth, fairly large and with an apical germ pore.

The majority grow on dung or on ground with a high organic nitrogenous content. Other habitats include tree stumps, rotten timber, damp carpets, walls etc. Some of the more substantial species are edible and excellent but *Coprinus atramentarius* should not be eaten with alcohol as this may cause nausea.

159 Coprinus picaceus (Fries) S. F. Gray
Magpie fungus

Cap: 4–8 cm high. Oval at first then campanulate. Covered by a white veil which, on bursting, leaves white felty patches on a black background.
Gills: Close and free. White at first, later pale brown and finally black.
Spores: Black and ovate. 13–17 × 10–12 μm.
Stipe: Tall and white with a bulbous base. No ring.
Flesh: Thin with an unpleasant smell.
Habitat and season: Grows singly in the rich soil of frondose woods. Is occasionally found from September to November.
Edibility: Yes, but only when young.

160 Coprinus micaceus Fries
Glistening coprinus

Cap: 2–4 cm. Ovate, then campanulate and finally expan-

157 Agaricus campestris

158 Agaricus silvicola

159 Coprinus picaceus

160 Coprinus micaceus

ded. Young specimens are ochre brown but greyish at margin and darker at disc. Strongly striate and covered with mica-like particles which soon disappear. Older specimens become date-brown and rimoso-sulcate. Auto-digestion is only slight.
Gills: Adnate, close and lanceolate. At first dirty white then through to dark brown.
Spores: Black and lemon-shaped. 7·5–12 × 6–7 × 4·5–6 μm.
Stipe: Thin, hollow and finely silky (lens). Dirty white.
Habitat and season: Usually densely tufted on old stumps and rotting timber. Often appearing repeatedly in the same spot and at any time of the year when suitable conditions prevail. Common May to December.
Edibility: Said to be edible with a pleasant taste.

* **Coprinus truncorum** (Schaeff.) Fries

Is similar to *C. micaceus* but has a smooth stipe, and is free of silky hairs.

* **Coprinus silvaticus** Peck

Also like *C. micaceus* but grows tufted on soil attached to buried wood under trees. The gills are crowded and brownish-black. Auto-digesting. Stipe is brownish and hairy (lens).

161 Coprinus trisporus Kemp & Watling

Cap: 5–17 mm high. 7–8 mm wide. Acorn-shaped, soon elliptical, then campanulate to expanded. White or whitish, somewhat grey or ochraceous at disc. Entirely floccose at first, but apart from centre becoming smoother with margin plicate-striate. Auto-digesting.
Gills: Crowded and free. Whitish then black. Edge white-floccose when young.
Spores: Black in mass, broadly elliptical, slightly amygdaliform in side view. 9–10(10·5) × (5·5)6–6·5 μm. Basidia 3-spored; which is characteristic.
Stipe: Sub-equal or slightly attenuated upwards. Fragile and hollow, finely pruinose at first, whitish.
Flesh: Whitish and thin. Odour distinctly nauseous.
Habitat and season: On dung from May to November.
Edibility: Worthless.

162 Coprinus plicatilis (Fries) Fries

Cap: 0·5–1·5 cm high. Ovate at first but soon more or less flat, 1·5–2 cm across. Pale brown becoming pale blue-grey with brownish disc and deeply sulcate. It could be likened to a miniature parasol. It is an extremely delicate and fragile species, soon splitting. Little or no auto-digestion.
Gills: Greyish-black and distant, adhering to a distinct collar.
Spores: Black in the mass, heart-shaped. 9·5–13 × 9–11 × 5–7 μm.
Stipe: White, smooth and equal. Very brittle.
Flesh: Virtually none.
Habitat and season: Fairly common in pastures, lawns and roadside verges. Growing solitary or a few together, easily overlooked. Found from May to November.
Edibility: Unsubstantial and worthless.

163 Coprinus atramentarius (Fries) Fries
Common ink cap

Cap: 3–7 cm. At first ovate, then bell-shaped. Grey to brown in colour. Fleshy and scaly at centre. Radially ribbed and often cracking to expose greyish flesh beneath.
Gills: Crowded, broad and free. Whitish but soon blackening and dissolving into an inky fluid (auto-digestion).
Spores: Smooth, black and elliptical. 8–10 × 4–5 μm.
Stipe: Whitish and smooth. Tapering upwards and with a ring-like zone at base.
Flesh: Greyish-white and thin.
Habitat and season: Grows in tight clusters on the ground in open woods, gardens and fields. Often near to stumps or woody plants to which they are probably attached. May to November.
Edibility: Good *before* auto-digestion commences. It must be pointed out that alcohol *should not* be taken with this fungus. Under such circumstances it is known to cause sickness.

164 Coprinus comatus (Fries) S. F. Gray
Shaggy cap or Lawyer's wig

Cap: 4–12 cm high and cylindrical. White and giving the appearance of being tiled with shaggy white scales. Soon becoming brownish towards the centre. Later the cap expands from below with the margin turning upwards and splitting.
Gills: White at first and closed as the pages of a book, thin and very broad. Later less close, becoming pink then black and finally auto-digesting.
Spores: Black and elliptic. 12–14 × 7–9 μm.
Stipe: Erect, slender, hollow and white with a thin movable ring. Bulbous at base.
Flesh: Thin and watery.
Habitat and season: Can be found in fields, by roadsides, also on rubbish tips. It likes a loose sandy soil. Often growing in groups or clusters. Common from May through to November.
Edibility: Take young specimens only and eat on day of collection, before auto-digestion takes place. Has a delicate flavour and very good.

* **Coprinus cinereus** (Fries) S. F. Gray
Coprinus fimetaris (Linn.) Fries
Coprinus macrorhizus (Fries) Rea
Dung-heap ink-cap

Cap: 2·5–5 cm across when expanded and 1–3 cm high. Oval or campanulate, then soon conico-expanded. Later splitting radially before auto-digesting. At first it is covered with a dense dirty-white woolly coating which breaks up into patches and eventually falls away leaving the cap shiny, the centre brownish-grey and the margin dark grey and striate.
Gills: Crowded and free. White but very quickly dissolving into a black liquid.
Spores: Violaceous black, elliptical and smooth with a germ pore. 10–12 × 6–6·5 μm.
Stipe: 2·5 to 8 cm long. Fragile and attenuated upwards. Has a taproot-like base which penetrates the substrate. White and covered with similarly coloured scales.
Habitat and season: Common on manure heaps or on very rich soil, at any time of the year.
Edibility: Not known.

* **Coprinus niveus** (Fries) Fries
Snow-white coprinus

Cap: 1–2 cm across and 1–3 cm high. Fragile, ovate then campanulate with margin split and upturned. Pure white and when fresh has a dense covering of chalky-white meal.
Gills: Crowded, narrow and black. Soon auto-digesting with cap.
Spores: Black and lemon-shaped. 12–18 × 10–12 × 8–10 μm.

161 Coprinus trisporus

162 Coprinus plicatilis

163 Coprinus atramentarius

164 Coprinus comatus

Stipe: Up to 8 cm long. Slightly tapering upwards. White with evanescent woolly floccules. Hollow and very fragile.
Flesh: White and unsubstantial.
Habitat and season: Common on cow and horse dung, singly or gregariously. May to November.
Edibility: Not known.

* **Coprinus narcotinus** (Fries) Fries

Similar to *C. niveus* but is greyish with a strong unpleasant odour. The spores have a colourless outer wall.

* **Coprinus stercorarius** (Bulliard) Fries

Similar to *C. niveus* but grey in colour. Springs from a small black sclerotium.

Genus-*Psathyrella*

Medium to small, fragile species. Centrally stipitate and with thin, regular, membranous cap. The gills are adnate or free. A ring may be present or not. Spores are dark violaceous-brown to black in the mass, smooth and with germ pore. Habitats include grassland, roadsides and on or near stumps of trees etc. A few are edible, others are bitter and worthless.

165 Psathyrella gracilis (Fries) Quélet

Cap: 1–3 cm. Conical to convex and often wrinkled or striate, no veil. Pale tan colour when dry, but brown when moist. Tinged pink with age.
Gills: Broadly adnate and rather distant. Grey to black edged with pale rose.
Spores: Purplish-black, elliptic. 11–13 × 5·5–6·5 μm.
Stipe: Long, slender and pallid with a short fibrillose tap-root.
Flesh: Greyish-white and very thin.
Habitat and season: Very common in frondose woods and hedge banks where it may be found growing in tufts from August to November.
Edibility: Unsubstantial and worthless.

166 Psathyrella hydrophila (Mérat) Maire
Hypholoma hydrophilum (Mérat) Quélet

Cap: 2–6 cm. Convex then expanded. With conspicuous white veil when young, the remains of which are often found adhering to cap margin. Fragile, smooth and hygrophanous. Deep date-brown when moist, drying out fawn.
Gills: Crowded and adnate. Pale at first then dark brown.
Spores: Dark purplish-brown. 5–7 × 3–4 μm.
Stipe: White with a dark ring-like zone. Slender, fragile and usually curved.
Flesh: Thin and pale.
Habitat and season: Grows on or near old frondose tree stumps, densely tufted and overlapping. Common May to November.
Edibility: Bitter tasting and worthless.

* **Psathyrella spadicea** (Fries) Singer
Psilocybe spadicea (Fries) Quélet

Very similar to *P. hydrophila* but has no veil, hence no ring-like zone on stipe. Found very occasionally April to November.

* **Psathyrella candolleana** (Fries) Maire
Hypholoma candolleanum (Fries) Quélet
Drosophila candolleana (Fries) Quélet

Cap: 2–6 cm. Campanulato-convex then expanded with denticulate remains of veil at margin. Glabrous, pale yellow or whitish.
Gills: Crowded, narrow and adnexed, rounded behind. Greyish-lilac then brown, finally dark brown.
Spores: Dark purplish-brown and elliptical. 7–8 × 4–4·5 μm.
Stipe: Slender and fragile, often curved. Thickened at base.
Flesh: White and very thin. Odour mild, taste agreeable.
Habitat and season: Common and often densely caespitose on the ground or on the stumps of frondose trees. From May to November.
Edibility: Good but unsubstantial.

167 Lacrymaria velutina (Fries) Konrad & Maublanc
Hypholoma velutinum (Fries) Kummer
Psathyrella lacrymabunda (Fries) Moser
Weeping widow

Cap: 2–8 cm. More or less bell-shaped at first, then expanded but remaining somewhat umbonate. Young specimens are covered with woolly fibrils which continue below cap margin (the remains of white fibrillose veil). Later the cap becomes smooth and hygrophanous. Variable in colour, clay-brown, bright orange or tawny. Usually brightest at apex.
Gills: Adnexed or adnate and crowded. Dark purplish-brown and mottled with white edges. They exude minute droplets of water giving a glistening effect.
Spores: Black, warted and lemon-shaped. 8–11 × 5–6 μm.
Stipe: Slender and fragile, paler than cap especially at apex. Equal and fibrillose, the veil leaves a zone on the stipe which becomes covered when the dark spores begin to drop.
Habitat and season: Solitary or clustered especially in grass of roadside verges, edges of woodland paths and wastelands. May be found April to November.
Edibility: Edible and good but slightly bitter and reminiscent of Hypholomas.

Genus-*Panaeolus*

Mainly small species with slightly fleshy caps. Centrally stipitate with regular, campanulate or conical caps. The margin extends beyond the gills. They are viscid when moist and shining when dry. The gills are adnate, the face mottled when young due to uneven spore dispersal. Spores are blackish, smooth and lemon-shaped with apical germ pore. Rich pastures and dung are their normal habitats. None is considered esculent and if eaten can cause intoxication or delirium.

168 Panaeolus sphinctrinus (Fries) Quélet

Cap: 1·5–2·5 cm. Bell-shaped or parabolic, never expanded. When wet, translucent and smoky-black. In dry weather, opaque and paler, more greyish and silky.

165 **Psathyrella gracilis**

166 **Psathyrella hydrophila**

167 **Lacrymaria velutina**

168 **Panaeolus sphinctrinus**

Margin at first white-dentate from fragments of veil (see illustration).
Gills: Adnate, ascending, crowded and broad. Greyish-black and mottled but edge may be white.
Spores: Black, lemon-shaped and smooth. 14–15 × 9–10 μm.

Stipe: Long, slender, straight, fragile and hollow. Smoky-grey and appearing powdery except at apex.
Flesh: Reddish.
Habitat and season: Common on dung or rich ground. From May to November.
Edibility: No.

169 Panaeolus semiovatus (Fries) Lundell POISONOUS
Anellaria separata (Fries) Karsten

Cap: 1–6 cm. Distinctly semi-ovate. Greyish to pale tan, viscid when wet, shining as if polished when dry. Often crinkled and cracked. The margin extends beyond gills and is often ornamented with veil fragments.
Gills: Adnate, broad and crowded. At first greyish mottled but later turning black.
Spores: Black, ovate, elliptic. 16–20 × 9–12 μm.
Stipe: Straight and usually slender, easily broken. Has a greyish-white ring, the remnants of which often adhere to the surface. The general coloration is similar to cap but usually darker towards the slightly swollen base.
Flesh: White and very thin.
Habitat and season: Very common on dung and rich garden soil from May to November.
Edibility: Said to be *poisonous.*

* **Panaeolus papilionaceus** (Fries) Quélet POISONOUS

Cap: 2–3 cm. Hemispherical to convex, often cracking. Margin extends over the gills. Greyish-white with yellowish tinge at disc in moist conditions. When dry it is yellowish with a reddish-brown tinge at disc.
Gills: Crowded, adnate and broad. Blackish and spotted, usually with a whitish edge.
Spores: Black and lemon-shaped. 14–16(18) × 10–12 × 8–9 μm.
Stipe: Slender and equal or slightly attenuated from a thickened base, hollow. Below it is wavy or curved and whitish or slightly pink.
Flesh: Thin and pallid. Mild odour and taste.
Habitat and season: Widespread and fairly common locally, grows in pastures, meadows, lawns and grassy places in open woodlands. From June to October.
Edibility: *Poisonous.*

Genus-*Stropharia*

Large to small-sized species. The cap is somewhat fleshy, often slimy and with separable pellicle. The flesh is continuous with that of the central stipe which has a distinct membranous veil ring. The gills are adnate and pale brown, becoming dark brown or purplish. The spores are dark brown to purple-brown, smooth and with germ pore.

Found mainly on dung or in grass. Some species are considered esculent but authorities differ on this point; we suggest they are better left alone.

170 Stropharia aeruginosa (Fries) Quélet POISONOUS
Geophila aeruginosa (Fries) Quélet
Verdigris agaric

Cap: 2–7 cm. Convex with a central boss. Green to deep turquoise but yellowing with age. Slimy, but this is usually washed away by rain. Floccose at first and margin festooned with whitish scales.
Gills: Broad, adnate and crowded. Lilac but later becoming violet brown.
Spores: Blackish, tinted violet under microscope, smooth with germ pore. 7·5–9 × 4·5 μm.
Stipe: Hollow and equal. Slimy at first and similar in colour to cap. Floccose below the ring and smooth above.
Ring: Frayed and white, becoming stained blackish by ripe spores. As the ring is set high, it is not visible in our illustration.
Flesh: Thin. White with greenish tinge. Smell resembles that of radish.
Habitat and season: In grass e.g. lawns and pastures; often well hidden. Fairly common from June to November.
Edibility: No. Disgusting flavour and said to be *poisonous.*

Larger than *S. aeruginosa* is *Stropharia ambigua* which has a yellow-brown cap. A North American species found in both spring and autumn. *Not edible.*

171 Stropharia aurantiaca (Cooke) P. D. Orton

Cap: 1·5–5·5 cm. Convex then expanded convex, margin stays incurved for some time. Very viscid when moist. Wrinkled round the disc, margin with whitish dentate scales at first. Colour: red-lead or crimson-lake to deep wine-red, sometimes paler at margin.
Gills: Adnate or slightly emarginate, rather crowded. Whitish, then pale olivaceous or brownish-olive. Gill edge white floccose at first, later sometimes with reddish stains.
Spores: Purplish-brown in mass, elliptical with germ-pore. 11–13 × 6–7·5 μm. Basidia 4-spored.
Stipe: Slender and equal with slightly thickened base. Whitish then pale yellowish or ochraceous.
Flesh: Concolorous in cap. Yellowish or ochraceous in stipe. Inodorous.
Habitat and season: An uncommon species which grows in grass or on sawdust. From October to November. Not listed for N. America.
Edibility: No.

* **Stropharia hornemannii** (Fries) Lundell & Nannfeldt
Naematoloma hornemannii (Fries) Singer POISONOUS

Cap: 5–10(15) cm. Convex then expanded, viscid. Straw-colour to chestnut-brown, often with a violet tinge when young.
Gills: Crowded, broad and adnate. Pallid then greyish-violet.
Spores: Purple-brown and elliptic. 11–14 × 6–8 μm.
Stipe: Stout. Straw-colour with evanescent whitish scales.
Ring: Ample and fleshy.
Flesh: Pallid. Odour disagreeable.
Habitat and season: A rare species found on fallen

169 Panaeolus semiovatus

170 Stropharia aeruginosa

171 Stropharia aurantiaca

172 Stropharia semiglobata

branches and wood chips in conifer woods. From September to November.
Edibility: *Poisonous.*

172 Stropharia semiglobata (Fries) Quélet
Stropharia stercoraria (Fries) Quélet
Dung round-head

Cap: 1–4 cm. Very variable in size. Long remaining hemispherical then somewhat expanded. Viscid, shining when dry. Colour straw-yellow.
Gills: Crowded, very broad and adnate. Pale purplish clouded with black, then brown.
Spores: Dark brown. Smooth, large and elliptical 17–20 × 9–10 μm.
Stipe: Long in proportion to size of cap. Slender, straight, smooth and hollow. Slimy below ring-like zone. Whitish-yellow but paler at apex.
Ring: Viscid and thin, often only represented by a dark zone.
Flesh: Pale and thin.
Habitat and season: Very common growing on dung, either solitary or gregarious. From April to November.
Edibility: Suspicious.

Genus-*Hypholoma*

Structurally the same as Tricholoma but with purplish-black spores. In young specimens, the cap margin is incurved and often with persistent velar remains.

Found mainly densely tufted on wood.

173 Hypholoma fasciculare (Fries) Kummer
Naematoloma fasciculare (Fries) Karsten
Sulphur tuft or Clustered woodlover

Cap: 2–7 cm. Convex then plane, subumbonate or obtuse, smooth. Margin incurved at first and often with velar remains. Sulphur-yellow but usually orangy at the umbo.
Gills: Adnate and very crowded. At first greenish-yellow, then olive-green, finally tinged purple-black by the spores.
Spores: Purplish-black, elliptical or oval. 5–7 × 3·5–4·5 μm.
Stipe: Slender and equal, hollow and incurved. Concolorous with cap, usually with a faint ring-like zone, darker below with squamules.
Flesh: Thin and yellow. Bitter taste.
Habitat and season: Very common, grows in dense tufts on frondose, and occasionally coniferous, stumps and sickly trees etc. From January to December but chiefly in the autumn.
Edibility: Inedible. Possibly *poisonous.*

* **Hypholoma sublateritium** (Fries) Quélet

Cap: 4–9 cm. Convex then plane, fleshy and smooth. Orange or brick-red but paler at margin and often with velar remains.
Gills: Crowded and narrow, adnate or sinuato-adnate. Yellow at first but finally greyish-black to dark brown.
Spores: Purplish-black, elliptical or oval. 6–7 × 3–4·5 μm.
Stipe: Stout and often curved. Attenuated downwards. Yellowish above and reddish-brown below. Scaly fibrillose, often with a ring-like zone.
Flesh: Firm and yellow but rusty at base of stipe. Taste may or may not be bitter.
Habitat and season: Fairly common growing densely tufted on stumps etc. of frondose trees. From August to November.
Edibility: Not known. Suspect.

* **Hypholoma capnoides** (Fries) Kummer
Naematoloma capnoides (Fries) Karsten

Cap: 2–6 cm. Ochre yellow.
Gills: Whitish or bluish-grey at first, then fuscous purple.
Spores: Purplish-black and ovate. 7–9 × 4–5 μm.
Stipe: Pallid with whitish apex.
Flesh: Pale yellow.
Habitat and season: Usually densely tufted on coniferous stumps. From spring to autumn.
Edibility: Said to be edible.

Genus-*Psilocybe*

Small species. The cap is moist, fleshy and regular with incurved margin when young. Gills are adnate or adnexed, fuscous, brownish or purplish. The stipe is central and sub-cartilaginous. Spores are dark brown or purple-brown, smooth and elliptical. Found growing on soil, in grass or on dung.

174 Psilocybe semilanceata (Secretan) Kummer
Geophila semilanceata (Secretan) Quélet
Liberty cap

Cap: 1–2 cm across and up to 2 cm high. Acutely conical and distinctly cuspidate, never expanded. Pellicle separable and viscid in moist conditions. Yellowish or clay-colour when dry but soon becoming tinged with olive-green.
Gills: Crowded and narrow, adnate or adnexed. Creamy then purple-black.
Spores: Purple-brown and lemon-shaped with germ pore. 12–14 × 7–8 μm.
Stipe: Slender, equal and flexuose, internally pithy. Cap-colour but lighter at apex.
Flesh: White and thin.
Habitat and season: Common in pastures, lawns and verges etc. From August to November.
Edibility: Worthless; causes delirium.

The variety *caerulescens* (Cooke) Saccardo, has a blue-green base to the stipe.

* **Deconia coprophila** (Fries) Karsten
Psilocybe coprophila (Fries) Kummer

Cap: 1–2·5 cm. Hemispherical then expanded. Slightly viscid and shining. Fleshy and with a detachable pellicle. White and downy when very young but later umber to dark brown.
Gills: Broad, crowded and arcuate, rather decurrent. Livid-brown.
Spores: Purple-brown. 12–14 × 6·5–8 μm.
Stipe: Scaly at first and centrally pithy, then glabrous, shining and hollow. Attenuated upwards. Pale cap-colour.
Flesh: Has a mealy taste.
Habitat and season: An occasional species growing on dung or in grass from August to November.
Edibility: Worthless.

Genus-*Pholiota*

Medium to large-sized species. The cap is fleshy and regular, viscid or dry, squarrose or naked. The flesh is continuous with that of central stipe which usually bears a

173 Hypholoma fasciculare

174 Psilocybe semilanceata

175 Pholiota squarrosa

176 Conocybe lactea

membranous ring. The gills are adnate or decurrent by a tooth, pallid at first then brown. Spores are yellow-brown or rusty-brown, smooth and without a germpore. They grow on wood or on the ground and are often caespitose. Saprophytic or parasitic. Some are esculent.

Members of the genus Agrocybe have been placed in Pholiota.

175 Pholiota squarrosa (Fries) Kummer
Shaggy pholiota or Scaly cluster fungus

Cap: 2–10 cm across. At first convex with an inrolled margin, expanding later and becoming flattened. Straw-yellow with a covering of dry, dark brown, recurved scales. Centre of cap deeper in colour. Margin often fringed with remains of the partial veil.
Gills: Adnate and decurrent by a tooth. Straw-yellow at first but later rust-coloured.
Spores: Pale brown, oval. 6–8 × 3–4 μm.
Stipe: Covered with scales which are so recurved as to give the appearance of a series of rings, but smooth above the small frayed ring. Of a similar colour to cap but darker at base. Grows up to $12\frac{1}{2}$ cm in length.
Flesh: Yellowish-brown with a strong pungent smell.
Habitat and season: Can be found growing in tiered clusters from the base of frondose trees. September to November.
Edibility: Indigestible and not recommended.

* **Pholiota adiposa** (Fries) Kummer
Fat pholiota

Cap: 8–16 cm. Yellow to golden-yellow and covered with viscid rusty-brown scales.
Gills: At first straw-coloured, later becoming rusty-ochre.
Spores: Oval. 5–6 × 3–3·5 μm.
Stipe: Similarly beset with viscid scales but yellowish and smooth above ring zone.
Flesh: Pale yellow but shading to rusty-brown at base of stipe.
Habitat and season: Caespitose at the base of beech trees and stumps. From August to October.
Edibility: No.

* **Pholiota alnicola** (Fries) Singer
Flammula alnicola (Fries) Kummer

Cap: 3–8 cm. Convex then expanded, slightly viscid at first but later dry and smooth. Margin superficially silky when young. Yellow, becoming tawny or greenish-yellow.
Gills: Adnate and fairly crowded. Pale yellow, finally ferruginous.
Spores: Brown and ovate. 7·5–9 × 4–5·5 μm.
Stipe: Long and more or less equal, fibrillose, usually curved or wavy. Yellow above to ferruginous below.
Flesh: Yellow in cap, brown in base of stipe. Bitter taste but a very pleasant odour.
Habitat and season: Occasional, singly or a few together on frondose trees and stumps, especially birch. From September to November.
Edibility: Worthless.

* **Pholiota apicrea** Fries
Flammula apicrea (Fries) Gillet

A rare fungus. Very similar to *P. alnicola* but with mild tasting flesh.
Spores: Ovate. 8–10 × 4–5 μm.
Edibility: Worthless.

* **Pholiota flammans** (Fries) Kummer
Dryophila flammans (Fries) Quélet

Cap: 2–6 cm. Convex then plane and slightly umbonate, dry. Bright chrome lemon then tawny yellow with scattered recurved sulphur-yellow squamules. Margin involute at first.
Gills: Adnexed and crowded. Yellow then tawny.
Spores: Brown and oval. 4–5 × 2·5–3 μm.
Stipe: Equal and flexuous, often curved. Stuffed then hollow. Yellow and furnished with crowded squamules of bright yellow up to the ring.
Ring: Superior and entire, or reduced to scurfy scales. Yellow.
Flesh: Firm and yellow. Not hygrophanous.
Habitat and season: Occasional, singly or a few together on pine stumps or trunks. From August to October. A striking and beautiful species.
Edibility: No.

* **Pholiota subsquarrosa** (Fries) Quélet
Dryophila subsquarrosa (Fries) Quélet

Cap: 4–6(7) cm. Convex or campanulate then plane and gibbous. Viscid, reddish-brown with darker adpressed floccose scales.
Gills: Crowded and almost free. Yellow then dingy tan.
Spores: Brown and oval, 4·5–6 × 2·3 μm.
Stipe: Equal, stuffed then hollow. Yellow, furnished with darker scales up to the superior ill-defined ring but usually smooth above it.
Flesh: Tough and yellowish. Lacking odour.
Habitat and season: A rare fungus. Occasional on conifer stumps or on ground nearby. From September to November.
Edibility: Yes.

* **Pholiota mutabilis** (Schaeff. ex Fries) Kummer
Kuehneromyces mutabilis (Fries) Singer and A. H. Smith
Galerina mutabilis (Fries) P. D. Orton
Changing pholiota

Cap: 3–8 cm. Convex then plane and usually obtusely umbonate. Hygrophanous, date-brown when moist, drying out paler from disc.
Gills: Crowded, broad, thin and adnato-decurrent. Pallid then cinnamon.
Spores: Brown and elliptical. 7–8 × 4–5 μm.
Stipe: Slender, firm, equal and usually curved, stuffed then hollow. Pale brown above, dark brown below. Scaly up to ring zone at first.
Ring: Superior and membranous.
Flesh: Dirty white to brownish.
Habitat and season: Common and densely caespitose on or near stumps and fallen trunks of frondose trees. Spring to early winter.
Edibility: Very good.

* **Pholiota destruens** (Brondeau) Gillet
Pholiota heteroclita (Fries) Quélet
Destructive pholiota

Cap: 6–10 cm. Convex then expanded and very obtuse. Compact and hard when young. Often eccentric, margin incurved and shaggy. Brown beset with white cottony scales.
Gills: Crowded, slightly adnexed, very broad and rounded behind. Pallid then brown.
Spores: Cigar-brown and oval. 7·5–8·5 μm.
Stipe: Short, very stout and solid, base bulbous and rooting. Scaly like cap.

Ring: Apical, tomentose and floccose.
Flesh: Thick and firm. White in cap, brown with age in stipe. Odour reminiscent of horse radish, taste bitter.
Habitat and season: Rare. Found singly or a few together on stumps of dead frondose trees, rarely on living trees. Causes a heart-rot. From August to November.
Edibility: No.

* **Pholiota aurivella** (Fries) Kummer
Golden pholiota

Cap: 5–12 cm. Campanulate then convex, compact at the disc, gibbous when expanded. Viscid, deep yellow, ferruginous towards centre with darker adpressed spot-like scales. Margin involute and floccose when young.
Gills: Crowded, broad and adnate. Pale yellow then ferruginous.
Spores: Fawn to smoky-brown, oblong and smooth. 8–9 × 4–6 μm.
Stipe: Dry, equal and usually curved. Yellow above, rusty-brown below, covered at first with ferruginous fibrillose scales below ring.
Ring: Fibrillose and soon disappearing.
Flesh: Pale yellow, reddish-brown in stipe base.
Habitat and season: Occasional. Tufted in knot-holes, and cracks of old stumps etc. of frondose trees. September to November.
Edibility: Unknown.

* **Agrocybe cylindracea** (DC ex Fries) Maire
Agrocybe aegerita (Briganti) Singer
Pholiota aegerita (Briganti) Quélet

Cap: 3–14 cm. Convex then plane, finally centrally depressed. Viscid in moist patches at disc. Margin incurved and scalloped. Pale tan, paling to whitish at the margin.
Gills: Crowded, broad, thin and decurrent by a tooth. Whitish then ochraceous.
Spores: Deep brown, elliptical and smooth. 8–11 × 5–6·5 μm.
Stipe: Usually slender. White, discolouring yellow or brownish below. Fibrillose and usually striate.
Ring: Superior, white, reflexed and membranous.
Flesh: Tender and white. Brownish under cap cuticle and stipe base. Odour pleasant and with a taste of hazelnuts.
Habitat and season: Caespitose on stumps and old trunks, especially of elder, poplar and elm. May be found at any time of year.
Edibility: Excellent and with a delicate flavour, a much sought after esculent. Also cultivated in parts of Europe on discs of poplar wood which are covered with a light layer of soil after being rubbed with the gills of the fungus.

Genus-*Conocybe*

Small species. Cap regular with a central stipe. The gills are yellowish to cinnamon. Spores are yellow to brown with a germ pore. They are found growing mainly in grass. Not really esculent.

A closely related genus. Agrocybe differs in the duller coloured gills and spores.

176 Conocybe lactea (J. Lange) Métrod
Galera lactea J. Lange

Cap: 1–3 cm high and up to 1·5 cm across at margin. Cylindrical then campanulate. Wrinkly striate when moist. smooth when dry. Creamy to pale tan.
Gills: Adnexed, narrow and moderately crowded. Cinnamon colour.
Spores: Yellowish-brown and elliptic with germ pore. 11–14 × 6–8 μm.
Stipe: Slender and rigid but extremely fragile. Narrowing slightly upwards from a small but definite bulbous base. Whitish and hollow.
Flesh: Thin and pallid.
Habitat and season: Common but often overlooked. In pastures, roadside verges and sand-dunes. Usually only in ones or twos. July to October.
Edibility: Worthless.

* **Conocybe tenera** (Fries) Kühner
Galera tenera (Fries) Quélet

Cap: 1–3 cm. Conical, thin and fragile. Smooth, hygrophanous when moist and can appear striate. Ochre-brown or cinnamon, more yellowish when dry.
Gills: Cinnamon. Adnate then free, crowded and ascending.
Spores: Yellow-brown, smooth and oval, with germ pore. 10·5–12 × 5·5–6·5 μm.
Stipe: Long, thin and fragile. Narrowing upwards from a small bulb at the base. Similar in colour to cap.
Habitat and season: Very common in grass, especially lawns. From July to November.
Edibility: Worthless as an esculent.

* **Bolbitius vitellinus** (Fries) Fries
Yellow cow-pat toadstool

Cap: 1–5 cm across. Bright yellow, slimy and striate.
Gills: Cinnamon and free.
Spores: Rusty. 12·5–14·5 × 7–9 μm.
Stipe: Fragile, white to yellowish.
Habitat and season: A common species of rich pastures, lawns and dung. From July to November.
Edibility: No.

Genus-Inocybe

The genus is well-defined and most nearly related to Cortinarius. Easy to identify as a genus, but within this it is often specifically difficult or impossible to identify without recourse to a high-powered microscope. They are small to medium-sized and terrestrial.

The cap is dry and regular, convex or umbonate, radially fibrillose or scaly, sometimes with a slight cortina. The gills are usually crowded, rarely distant. Pallid at first but finally clay-coloured or with an olivaceous tinge. The spores are snuff-brown, but shape is of great importance in identification. The stipe is slender and central with neither ring nor volva but often with a bulbous base. The flesh is often with a distinctive smell—earthy, fruity, spermatic etc.

Most are mild tasting but probably *all are poisonous*; some are known to have caused deaths.

Subgenus-Inocybe
177 Inocybe posterula (Britz.) Saccardo
Inocybe descissa (Fries) Quélet

Cap: 2–5 cm. Campanulate with a prominent umbo. Rimose, and thickly downy. Pale fawn with a paler margin which is often fringed with remnants of the veil.
Gills: Crowded and more or less sinuate with a tooth. Whitish then clay colour edged with white.
Spores: Snuff-brown and almond-shaped. 8·5–10 × 4–5 μm.
Stipe: 5–7 cm. Whitish or cap colour. Powdery at apex.
Flesh: White. Smells faintly earthy.
Habitat and season: Very common in pine woods from August to November. Not recorded in America.
Edibility: Mild tasting but *should not be eaten* as *most* Inocybes are *poisonous* or *suspect.*

Subgenus-Inocybe
178 Inocybe geophylla (Fries) Kummer POISONOUS

Our illustration is of variety *lilacina* Gillet

Cap: 1–3 cm conical, but soon expanded with umbo. White with faint suggestion of yellow, silky-fibrillose. In the variety *lilacina* (as the illustration), it is of a lilac shade, often with a yellowish umbo.
Gills: Narrow, crowded and adnexed. White at first but soon becoming clay-coloured.
Spores: Snuff-brown and smooth. 9–11·5 × 5–7 μm.
Stipe: White, slender, smooth, equal but rarely straight. The variety *lilacina* differs only in colour, being lilac with an ochraceous base.
Flesh: White (lilac in var. *lilacina*), thin with an earthy or spermatic odour.
Habitat and season: Can be found commonly in deciduous woods, especially in damp shady places, often in grass. From June to November.
Edibility: *Poisonous.*

Subgenus-Inocybe
179 Inocybe fastigiata (Fries) Quélet VERY POISONOUS
Inocybe pseudofastigiata Rea

Cap: 3–10 cm. Conic-campanulate with acute or obtuse umbo. Margin incurved when young. Radially fibrillose and often cracked. Rarely adpressedly squamulose. Straw-yellow, browner later.
Gills: Crowded. Adnate, adnexed or free. Narrow, thick and ventricose. Yellow then olive with a white edge.
Spores: Snuff-brown, smooth and bean-shaped. 9–12 × 4·5–7 μm.
Stipe: 4–10 cm long and up to 1·5 cm thick. Solid, equal or slightly tapering at both ends. Silkily fibrous or flocculose, pallid and then brown. Apex white pruinose.
Flesh: White and fibrous, not rigid. Taste mild with a slightly bitter aftertaste.
Habitat and season: Common in frondose woodland especially beech. From July to November.
Edibility: *Very poisonous.*

Subgenus-Inocybe
* **Inocybe patouillardii** Bresadola DEADLY POISONOUS
Inocybe rubescens Gillet

Cap: 3–8 cm. Conical or campanulate then expanding but still umbonate, silky fibrillose. White or creamy then yellowish-brown, bruising brown and often becoming wholly red, with margin cracked and splitting.
Gills: Crowded, narrow, adnate to almost free. White then olive-brown, edge white floccose, bruising or becoming red.
Spores: Bistre, smooth and bean-shaped. 10–13 × 5·5–7 μm.
Stipe: 4–10 cm long by 1–2 cm thick. Firm and equal or with marginate bulb. White becoming red like other parts.
Flesh: White, thick and vinaceous tinged at disc and in base of stipe. Mild taste and insignificant odour.
Habitat and season: Common growing in troops on base rich substrate under frondose trees, especially beech and lime. From May to November.
Edibility: *Deadly poisonous.*

Subgenus-Inocybe
* **Inocybe godeyi** Gillet POISONOUS
Inocybe rickenii Kallenbach

Very similar in appearance to *I. patouillardii* but generally slightly smaller, it also becomes red with age.
Spores: Snuff-brown, smooth and amygdaloid. 9–11·5 × 5–7 μm.
For positive differentiation between the two species, gill cystidia must be examined under a microscope.

177 **Inocybe posterula**

178 **Inocybe geophylla**

179 **Inocybe fastigiata**

180 **Inocybe boltonii**

Subgenus-Clypeus

180 Inocybe boltonii Heim. PROBABLY POISONOUS

Astrosporina boltonii (Heim) Pearson

Inocybe subcarpta Kühner & Boursier

Cap: 2·5–5 cm. Campanulate with an obtuse umbo. Covered with dark umber radiating fibrils which are dense at the disc. Separating to expose buff flesh towards the margin.
Gills: Crowded. Greyish-brown then rusty-brown.
Spores: Snuff-brown. Oblong or subtriangular with blunt protruberances and a prominent apical knob. 8–12 × 5·5–7 μm.
Stipe: 3–6 cm long and up to 6 mm thick. Equal and fibrillosely striate. Light brown but white pruinose at apex.
Flesh: White in cap. Reddish-brown in stipe. Mild taste.
Habitat and season: Frequent in coniferous woodlands from July to November.
Edibility: *Probably poisonous.*

Subgenus Clypeus

181 Inocybe lanuginella (Schroeter apud Cohn) Konrad & Maublanc VERY POISONOUS

Cap: 2–4 cm. Conico-campanulate then expanded with a persistent umbo (somewhat Lepiota-like in appearance). Ochraceous-brown, darker at disc. Radiately fibrillose then separating into shaggy scales and exposing white flesh.
Gills: Rather wide, not very crowded and with intermediates. Cream then ochraceous.
Spores: Snuff-brown, oblong, nodulose with a pentagonal profile. 8–9(10) × 5–6 μm.
Stipe: Equal, slender and white, becoming ochraceous from base upwards. Fibrillose, apex pruinose.
Flesh: Whitish but slightly brownish in stipe. Mild taste and mealy odour.
Habitat and season: Gregarious in grass and under frondose or coniferous trees. September to November. Not recorded in America.
Edibility: Known to be *very poisonous.*

Genus-Hebeloma

Medium-sized species. The cap is regular and fleshy, smooth to sub-viscid. Veil present or not. Colour usually white to pale brown. Margin involute at first. The stipe is central and fleshy without a ring. The gills are sinuato-adnate, pallid to brown. The spores are lemon, almond or oval-shaped, yellowish to brownish and minutely rough. All species are terrestrial and often have a distinctive odour. None are esculent and some are poisonous.

182 Hebeloma crustuliniforme (St. Amans) Quélet POISONOUS TO SOME PEOPLE
Ring agaric or Poison pie

Cap: 3–15 cm. Campanulate becoming plane, slightly viscid. Margin involute and downy at first, veil absent. Pale watery-tan in colour.
Gills: Crowded, denticulate and sinuate. Exuding droplets in moist weather. Whitish at first, becoming brown.
Spores: Fuliginous-ochre and almond-shaped. 10–12·5 × 5–7 μm.
Stipe: Robust, firm and usually with a bulbous base. Whitish, and coarsely furfuraceous above.
Flesh: Whitish. Has an odour of turnip and tastes very bitter.
Habitat and season: Common in groups or rings under trees, hedgebanks, and gardens etc. Especially in damp situations. July to November.
Edibility: No and *poisonous* to some.

Genus-Naucoria

Small species with cap either convex or conical, then plane. The margin is incurved at first. The gills are adnate or free and pallid to brown. The spores are rusty-brown, large and rough. The stipe is central and cartilaginous. They grow mainly on the ground and quite often in damp areas under trees.

* **Naucoria escharoides** (Fries) Kummer
Alnicola escharoides (Fries) Romagnesi

Cap: 1–3 cm. Convex then expanded, scurfy, margin appears striate when moist. Pale yellowish at first but becoming cinnamon.
Gills: Crowded, adnate and narrow. Rusty-brown, edge paler and woolly.
Spores: Brown, elliptical and warted. 10·5–11·5 × 5·5–6·5 μm.
Stipe: Slender, equal and hollow. Fibrillose and fragile. Brown at base but paler above.
Flesh: Yellowish and thin.
Habitat and season: Gregarious under alders in damp places. Common from September to November.
Edibility: Worthless.

Genus-Cortinarius

A very large genus of small to large species, divided by authorities into six sub-genera, details of which will be found in more advanced works. The Cortinarius are easily recognised as such in the field, but are often very difficult to identify specifically.

The whole genus carries the following characteristics: the cap is regular. The flesh is continuous with that of the stipe. Cobweb-like veil (cortina) when young, which is evanescent. Gills in young specimens vary in colour but are *always* rust-coloured at maturity. The spores are yellowish to red-brown under the microscope but rust-brown in the mass, of various shapes and often ornamented.

They are typically ground-loving woodland species, often forming mycorrhiza with the trees under which they are found and with which they are associated. Some are considered esculent, but taken overall, knowledge is scant in this direction.

Subgenus-Cortinarius

183 Cortinarius pholideus (Fries) Fries
Inoloma pholideum (Fries) Ricken

Cap: 4–8 cm. Convex then expanded and somewhat umbonate, not viscid. Fawn or olivaceous-fawn and covered with dense erect darker squammules.
Gills: Crowded, thin and emarginate. Violaceous then cinnamon, becoming pallid with age.
Spores: Cinnamon and broadly elliptical. 7–8 × 4·5–6 μm.
Stipe: Slightly attenuated upwards. Silky-fibrous with numerous dark brown transverse squamules. Apex often tinged with lilac.

181 Inocybe lanuginella

182 Hebeloma crustuliniforme

183 Cortinarius pholideus

Flesh: Pale brown but violaceous at apex of stipe.
Habitat and season: Common in frondose woods especially birch bogs. From August to October.
Edibility: Uninteresting.

Subgenus-Sericeocybe
184 Cortinarius alboviolaceus (Fries) Fries
Inoloma alboviolaceum (Fries) Wünsche

Cap: 3–6 cm. Convex and broadly umbonate. Violaceous-white and innately silky.
Gills: Adnate, somewhat distant and serrulated. Light blue at first, later cinnamon.
Spores: Rusty-brown and ovate. 8–10 × 4·5–5 μm.
Stipe: Club-shaped. Violaceous-white but slightly bluish above.

184 Cortinarius alboviolaceus

Flesh: Juicy. Whitish to pale violet.
Habitat and season: Fairly common in frondose woods from August to October.
Edibility: Of no importance.

Subgenus-Cortinarius
* **Cortinarius violaceus** (Fries) Fries
Inoloma violaceum (Fries) Wünsche
Violet cortinarius

Cap is up to 12 cm. Convex then broadly rounded to almost flat. Both cap and stipe are of a uniform deep violet colour with an almost metallic appearance when dry. Found in all types of woodland but especially beech. One of the most beautiful and easily identifiable of all cortinarius.

Subgenus-Myxacium
185 Cortinarius trivialis J. Lange
Myxacium triviale (J. Lange) Moser apud Gams.

Cap: 5–10 cm. Convex then expanded, broadly umbonate. Clay-, date- or bay-brown and very glutinous. Pales at margin with age. Cortina whitish.
Gills: Adnate and crowded. Whitish or pale clay and finally cinnamon brown.
Spores: Rusty brown and almond-shaped. 10–13 × 6–7 μm.
Stipe: Fairly tall and stout. Spindle-shaped, rarely cylindrical. Apex whitish and striate. Lower regions covered with pale net-like scales on a cap-coloured background. Deeply rooting.
Flesh: Yellowish-white but brown under cap cuticle and lower part of stipe.
Habitat and season: Fairly common in deciduous woods especially in wet places under alder and willow. August to October.
Edibility: Unknown. Has a mild taste.

Subgenus-Myxacium
* **Cortinarius collinitus** (Fries) Fries

Somewhat similar to *C. trivialis* but the colours are brighter. The belt-like scales on the stipe are pale blue. Spores are 12–15 × 7–8·5 μm. Usually found under conifers.

Subgenus-Myxacium
186 Cortinarius pseudosalor J. Lange
Cortinarius mucifluoides (R. Henry) R. Henry

Cap: 3–7 cm. Conico-convex then slightly expanded. Viscid and somewhat umbonate. Smooth but maybe wrinkled at centre. Ochraceous-yellow or buff. Margin paler, smooth or slightly wrinkled-striate, sometimes tinged violet.
Gills: Adnate, distant when mature and often veined on the sides. Ochraceous clay then deep rust, edge paler or slightly violaceous.
Spores: Rusty brown and almond-shaped. 12–14 × 7–9 μm.
Stipe: Usually long, slightly spindle-shaped, often tinged violet. Apex silky striate, viscid below cortinal zone. Becoming shiny and scaly as it dries with age.
Flesh: Whitish, tinged yellow beneath umbo. Thick at disc.
Habitat and season: Fairly common in frondose woods especially beech. Very often mistaken for *C. elatior*. August to October.
Edibility: Not known. It has a mild taste and no odour.

Subgenus-Myxacium
* **Cortinarius elatior** Fries

Similar to but larger than *C. pseudosalor*. Cap 5–12 cm, strongly wrinkled and more darkly coloured. Spores: 12–15(17) × 7–8·5 μm. Found in all types of woodland especially beech. Much less common than *C. pseudosalor*.

Subgenus-Dermocybe
187 Cortinarius semisanguineus (Fries) Gillet

Cap: 2–5 cm across. Reddish-brown tinged with yellow, dry and umbonate.
Gills: Vermillion at first but later becoming rusty-brown. A younger specimen would display much redder gills than our illustration.
Spores: Brown, ovate. 6·5–8 × 4–5 μm.
Stipe: Slender and yellowish-brown, with fibrils of a rusty green.
Flesh: The same colour as stipe, yellowish-brown.
Habitat and season: Grows singly or in small groups pushing up through the carpet of pine needles in a coniferous wood. Also found in birch woods. Likes a sandy soil. Common from August to November.
Edibility: Not edible.

Subgenus-Dermocybe
* **Cortinarius sanguineus** (Fries) Fries
Blood-red cortinarius

Another non-edible mushroom to be found commonly in conifer woods during the late summer and autumn. Approximately the same size as the previous species. Cap, gills and stipe are of a blood-red colour, the spores are rusty brown. Has a slender fibrillose stipe.

Subgenus-Telamonia
188 Cortinarius glandicolor (Fries) Fries

Cap: 2–4 cm. Convex and depressed around an acute umbo. Normally dry, dark chestnut to blackish-brown, smooth. Margin with fine white fibrils when young, often splitting.
Gills: Broad, adnate, thick and distant. Cinnamon-brown.
Spores: Pip-shaped and finely rough. 8–9 × 4–5 μm.
Stipe: Slender and stuffed then hollow. Striately brown with darker fibrils and a white ring-like belt (veil remnant). Base slightly thicker and whitish.
Flesh: Very thin and concolorous with cap.
Habitat and season: Mainly a species of pine-woods but also found under frondose trees, usually a few together. Occasional from September to November.
Edibility: Unknown.

Subgenus-Telamonia
* **Cortinarius brunneus** (Fries) Fries

Similar to *C. glandicolor* but much larger. Cap is 7–10 cm and umber; dingy reddish-tan when dry. Spores are obliquely elliptical and granular. 10–12 × 6 μm.

185 Cortinarius trivialis

186 Cortinarius pseudosalor

187 Cortinarius semisanguineus

188 Cortinarius glandicolor

Subgenus-Telamonia

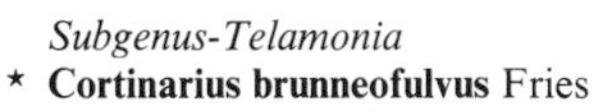

* **Cortinarius brunneofulvus** Fries

Allied to *C. brunneus* but cap tawny-cinnamon and larger than *C. glandicolor* being 4–6 cm across. Spores are elliptical and granular.

Subgenus-Telamonia
189 Cortinarius hinnuleus Fries

Cap: 3–7 cm. Convex then umbonate, somewhat conical. Smooth and fragile, often splitting from the edge inwards. Tawny or yellowish-brown.
Gills: Broad and adnate or sinuate. Distant when adult. Often connected by veins. Ochre yellow and then cinnamon.
Spores: Rust coloured and pip-shaped. 9–10 × 6–7 μm.
Stipe: Fibrillose, cylindrical and usually long. Yellowish-brown with a white ring-like zone above the middle. The veil is white and silky.
Flesh: Thin at cap margin and brownish. Reddish-brown in stipe.
Habitat and season: Common and usually gregarious. Grows on the ground in mixed woodlands. September to November.
Edibility: Unknown.

Subgenus-Telamonia
* **Cortinarius gentilis** (Fries) Fries

Similar to *C. hinnuleus* but more slender and brighter coloured. The veil forms one or more yellow zones on the stipe. Common in pine woods during autumn.

Subgenus-Telamonia
190 Cortinarius hemitrichus (Fries) Fries

Cap: 3–7 cm. Convex then expanded with a small umbo. Fuscous but brownish tan when dry. Densely covered (sometimes not so at centre) with superficial, whitish downy fibrils. Smooth later.
Gills: Adnate, thin and very crowded. Tan colour, then cinnamon.
Spores: Yellowish-brown, smooth and elliptical. 8·5–9 × 5·5 μm.
Stipe: More or less equal, hollow and fibrillose. Whitish at the apex, grey-brown below. Has a coating of whitish fibrils from and below the veil zone.
Flesh: Brownish. Thin except at umbo.
Habitat and season: Common, usually gregarious, on the grounds in woods, especially birch, also found on heaths. September to November.
Edibility: Not known.

Subgenus-Telamonia
191 Cortinarius armillatus (Fries) Fries

Cap: 5–10 cm. Campanulate then expanded. Fleshy at disc but thin elsewhere. Brick-red to brownish, paler at margin. Smooth or innately fibrillose. At first the margin is involute with adhering remnants of the reddish veil.
Gills: Very broad and moderately distant. Adnate, or ventricose behind and seeming to be adnexed. Pale then dark cinnamon.
Spores: Pale rust-brown. Almond-shaped and finely warted. 9–12 × 5–6 μm.
Stipe: Tall (up to 15 cm) and solid. Gradually attenuated upwards from a swollen or bulbous base. Pale greyish-brown and fibrillose with oblique zone or zones of cinnabar-red.
Cortina: Reddish-white and fibrillose.
Flesh: Pale tan.
Habitat and season: Fairly common on the ground in woodlands, especially birch in heathy areas. From August to October. A most striking and easily recognised species.
Edibility: Not known.

Subgenus-Phlegmacium
* **Cortinarius caesiocyaneus** Britzelmayr

Cap: 6–8 cm. Convex then plane, viscid. Pale blue when young, brownish with age.
Gills: Crowded. Whitish then bluish-violet and finally cinnamon.
Spores: Cinnamon and almond-shaped. 8–10 × 4–5 μm.
Cortina: Violaceous-white.
Stipe: Short and thick with a large bulb at base. Blue becoming ochraceous.
Flesh: Bluish in cap and stipe, yellowish in bulb.
Habitat and season: Fairly common in leaf debris of beech woods, especially in base rich areas. August to October.
Edibility: Unknown.

192 Rozites caperatus (Fries) Karsten
Pholiota caperata (Fries) Kummer
Gypsy mushroom

Cap: 5–15 cm. Ovate then campanulate and finally expanded. Moist, viscid in wet weather. Yellow or ochre, covered with superficial white flocci which are dense at the disc, becoming naked.
Gills: Crowded, adnate, narrow near the stipe, margin finely toothed. Yellowish then ochraceous.
Spores: Ochraceous, elliptical and finely warty. 11–14 × 7–9 μm.
Stipe: Thick, solid and fibrous. Equal or slightly swollen at base and with inferior volva. White but apex shining white. Squamulose from the superior deflexed ring.
Ring: Soft and white with yellowish or amethystine zone. Eroding with age.
Flesh: Thin and yellowish, soft and fragile. With a mild odour and a pleasant taste.
Habitat and season: Solitary or a few together on the ground in pine woods. Occasional from August to December.
Edibility: Good; can be dried or preserved in oil.

189 Cortinarius hinnuleus

190 Cortinarius hemitrichus

191 Cortinarius armillatus

192 Rozites caperatus

193 Gymnopilus junonius (Fries) P. D. Orton
Pholiota spectabilis (Fries) Kummer

Cap: 5–13 cm. Hemispherical at first and then convex. It is fleshy, dry and fibrillose. Tawny or golden yellow.
Gills: Adnate with a decurrent tooth, crowded and narrow. Yellow then ferruginous.
Spores: An ochraceous rusty brown and ovate. 8–10 × 5–6 μm.
Stipe: Solid, hard, fibrillose and ventricose. Sulphur-yellow. Mealy above the ring which is superior.
Flesh: Thick, hard and sulphur-yellow.
Habitat and season: Grows in tufts at the base of frondose and coniferous trees, also on stumps. A most striking and beautiful fungus.
Edibility: Not recommended, it is bitter and aromatic.

* **Gymnopilus penetrans** (Fries) Murrill
Flammula penetrans (Fries) Quélet

The cap is 3–6 cm and more or less umbonate. Golden to rusty-brown with no ring on stipe. Found predominantly in pine woods on logs and sticks etc. Fairly common August to November.

Genus-*Galerina*

Small thin species which are more or less conical. Their gills may be adnate or adnexed. The stipe is slender, central and cartilaginous. The spores are smooth. Ochreous to ferruginous in colour, often found growing in mossy places.

194 Galerina paludosa (Fries) Kühner
Galera paludosa (Fries) Kummer

Cap: Convex to umbonate with a very prominent papilla. Smooth, silky and brownish-yellow.
Gills: Adnate and broad. Greyish to brownish-yellow.
Spores: Ferruginous and smooth with a germ pore. 9–11 × 6–6·5 μm.
Stipe: Long and flexuose with a membranous ring which soon erodes. Floccose above. Colour similar to cap.
Flesh: Watery and ochraceous.
Habitat and season: Frequently found amongst Sphagnum on marshy ground. April to November.
Edibility: Worthless.

* **Galerina hypnorum** (Fries) Kühner
Galera hypnorum (Fries) Kummer

Much smaller than *G. paludosa*. Cap is 0·5 cm. Hemispherical or campanulate. Sand-coloured. Striate almost to the centre. Gills are adnate-emarginate. Tawny then rust-coloured.

* **Galerina mycenopsis** (Fries) Kühner
Galera mycenopsis (Fries) Quélet

Very similar to *G. hypnorum* but cap slightly larger 0·5–1·5 cm. When young has white fibrils on stipe.

Genus-*Crepidotus*

Mainly small or minute species. The cap is eccentric, lateral or resupinate. Usually thin with soft flesh and bracket-shaped like *Pleurotus*. Typically with brown spores. May be found growing on decaying wood, leaves and moss.

195 Crepidotus mollis (Fries) Kummer
Soft slipper toadstool

Cap: 1–7 cm. Shell or kidney-shaped, horizontal. Pale brownish.
Gills: Crowded and decurrent to a central point. Whitish, then watery cinnamon, often spotted.
Spores: Yellow-brown, smooth and elliptical. 6·5–8 × 4·5–6 μm.
Stipe: Rudimentary and lateral.
Flesh: Whitish, soft and gelatinous with a mild taste.
Habitat and season: Often growing tiered on stumps and decaying branches of frondose trees. Common from May to November.
Edibility: Yes.

* **Crepidotus calolepsis** (Fries) Karsten

Similar to *C. mollis* in habit but much smaller. Cap is 2 cm across with rufescent scales and firm, not gelatinous, flesh.

Genera-*Entoloma, Nolanea, Leptonia* and *Eccilia*

Sometimes referred together under Rhodophyllus. A large group with angular or wavy nodulose, deep salmon-pink spores which colour the gills at maturity. The gills are variously attached. They are typically terrestrial and centrally stipitate.

For exact determination of many species, the use of a microscope is essential.

Many species of this group are poisonous or suspect and care should always be taken.

Clitopilus is thought by some authorities to be related.

196 Entoloma aprile (Britz.) Saccardo
Rhodophyllus aprilis (Britz.) Romagnesi

Cap: 3–6 cm. Convex, then expanded with umbo. Often wavy and uneven. Silky greyish-brown when dry, dull date brown in moist conditions.
Gills: Distant, broad and emarginate. Greyish at first then pink.
Spores: Salmon pink. 8–10 × 7–8 μm.
Stipe: Tall and striate, becoming hollow. White to pale grey.

193 Gymnopilus junonius

194 Galerina paludosa

195 Crepidotus mollis

196 Entoloma aprile

Flesh: Firm. Pale grey.
Habitat and season: A vernal species appearing under frondose trees and shrubs during April and May. Not common.
Edibility: *Suspect.* Best avoided.

* **Entoloma sinuatum** (Fries) Kummer POISONOUS
Rhodophyllus sinuatus (Fries) Singer
Entoloma lividum (St. Amans) Quélet

Cap: 6–20 cm. Convex then plane, often umbonate or somewhat gibbous, smooth or slightly viscid. Margin usually splitting. Cinereous or ochraceous with leaden shades.

Gills: Broad and sub-distant with intermediates, nearly free. Yellowish at first then pinkish or ochraceous.
Spores: Rusty-pink and angular. 9–11 × 8–9 μm.
Stipe: Thick and striate, sometimes eccentric and curved. White with apex granular or powdery, becoming yellowish with age.
Flesh: Thick, firm and white. When young has an odour of new meal.
Habitat and season: Grows in groups or rings under frondose (rarely coniferous) trees. Fairly common from June to November.
Edibility: *Poisonous.*

197 Leptonia sericella (Fries) Barbier
Rhodophyllus sericellus (Fries) Quélet
Entoloma sericellum (Fries) Kummer
Agaricus molluscus Lasch

Cap: 1–2·5 cm. Convex or campanulate, then expanded or depressed. Smooth or slightly squarrose. Margin incurved and downy. White or pale ochre, often darker at disc.
Gills: Adnate or adnato-decurrent by a tooth. White then flesh-colour.
Spores: Pink and oblong. 9–11 × 6·5–7·5 μm.
Stipe: Slender and equal. Translucent and whitish.
Flesh: Thin and fragile with an earthy odour.
Habitat and season: Grows in grass and open woodland. Common from July to November.
Edibility: No.

* **Leptonia lampropus** (Fries) Quélet POISONOUS
Rhodophyllus lampropus (Fries) Quélet

Cap: 1–3 cm. Convex then expanded, minutely scaly or fibrillose. Grey to brownish.
Gills: Crowded, broad and adnate then free. White at first, pale pink later.
Spores: Pink and angular, 9·5–11·5 × 6·5–7 μm.
Stipe: Slender, smooth and cartilaginous. Deep blue but pale at base.
Flesh: Thin and bluish.
Habitat and season: Found in grassy places and on heaths. Fairly common from July to November.
Edibility: *Poisonous.*

198 Leptonia corvina (Kühner) P. D. Orton
Rhodophyllus corvinus Kühner

Cap: 1·5–5·5 cm. Hemispherical then campanulato-convex to expanded, centrally depressed. Margin at first incurved. Never striate. When young black, tinged bluish-violet to slaty-violet; finally more or less brownish-violet, and entirely flocculose to tomento-fibrillose.
Gills: Adnate and often more or less sinuate, never clearly decurrent, fairly crowded. At first clear white to creamy-white sometimes with a bluish tinge or again faintly tinted greyish-blue to violet at the edge, finally salmon-pink. Tomentose to flocculose under lens.
Spores: Pink and angular. 8·5–12 × 6·5–7 μm.
Stipe: Equal and furrowed. Central or not. Steel-grey, greyish-violet, dull to greyish-blue, becoming paler with age and whitish. Clearly tomentose, almost granularly floccose. With age particularly fibrillosely-striate, internally pithy or hollow.
Flesh: Thin. Odour and taste insignificant.
Habitat and season: Occasional in grass, lawns etc., also mossy areas in woodlands. October to November. Not recorded in America.
Edibility: Not known.

* **Nolanea cetrata** (Fries) Kummer PROBABLY POISONOUS
Rhodophyllus cetratus (Fries) Quélet

Cap: 2–4 cm. Broadly conical to convex then plane to somewhat depressed. Translucent and striate to the centre when moist, this characteristic is not very evident when dry. Yellowish-brown or tan.
Gills: Almost free, distant, broad and narrowed at each end. Pale yellowish becoming darker with age.
Spores: Pink, angular and oblong. 10–13 × 6–7 μm. 2-spored basidia.
Stipe: Up to 8 cm long, slender and striate. Pale yellow or cap colour.
Flesh: Brownish, watery and fragile. Odour indistinct.
Habitat and season: Common on the ground in coniferous woods, especially in damp places. September to November.
Edibility: Probably *poisonous.*

* **Nolanea icterina** (Fries) Kummer
Rhodophyllus icterinus (Fries) Quélet

Cap: 1–3·5 cm. Campanulate then convex, margin slightly striate, often becoming entirely reflexed and uneven. Hygrophanous and opaque, olivaceous buff to honey in wet weather but paler and silky when dry.
Gills: Adnexed or free, seceding and distant. Pallid becoming saffron with age.
Spores: Pink, oblong and nodulose. 9–12 × 6–8 μm.
Stipe: Up to 5 cm long. Stuffed and rigid, often compressed. Concolorous with cap or brownish. Often powdery but sometimes only at apex.
Flesh: Fragile. Odour of wintergreen or of boiled sweets.
Habitat and season: Occasional, solitary or caespitose on the ground in damp places. August to November.
Edibility: Worthless and suspect.

* **Nolanea staurospora** Bresadola
Rhodophyllus staurosporus (Bres.) J. Lange
Nolanea pascua (Fries) Kummer

Cap: 3–5 cm. Broadly conical or campanulate then expanded. Hygrophanous; date-brown and striate when moist; yellowish to pale cinnamon-brown, silky and non-striate when dry.
Gills: Almost free, crowded, broad and narrowed at each end. Off-white becoming flesh coloured.
Spores: Pink and quadrangular stellate. 9–10 × 7–9 μm.
Stipe: Long and slender, striate and hollow. Slightly paler than cap, white tomentose at the base. Very fragile.
Flesh: Brown and watery. Odour and taste indistinctive.
Habitat and season: Very common in pastures and clearings in woods. May to November.
Edibility: Suspect.

* **Leptonia babingtonii** (Bloxam) P. D. Orton
Nolanea babingtonii (Bloxam apud Berk. & Br.) Saccardo

197 **Leptonia sericella**

198 **Leptonia corvina**

199 **Paxillus atrotomentosus**

200 **Paxillus involutus**

Leptonia dysthales (Peck) Konrad & Maubl.
Inocybe bucknallii Massee

Cap: 5–15 mm. Very thin, conic to campanulate. Pale grey to sepia or greyish-brown. Adorned with shining-silky grey brown hairs.
Gills: Distant and adnate. Greyish-pink.
Spores: Greyish-pink, wavy, angular and long. 14–20 × 7–9 μm.
Stipe: Hollow and often wavy. Silvery-grey to grey-sepia and silky. With grey brown hairs.
Flesh: Very thin. Greyish.
Habitat and season: On bare ground in autumn. Do not know if it has been recorded in America.
Edibility: Worthless.

Note: *Leptonia babingtonii* must have grey-brown hairs on cap and stipe.

* **Eccilia sericeonitidia** P. D. Orton
Clitopilus undatus (Fries) Gillet

Cap: 2–4 cm. Convex then umbilicate or infundibuliform, wavy. Silky greyish-brown, often zonate. Margin incurved.
Gills: Deeply decurrent and narrow. Grey then somewhat flesh coloured.
Spores: Pink, angular and elliptical. 9–10 × 5–7·5 μm.
Stipe: Fibrous and usually curved, stuffed then hollow. Grey and mealy with whitish cottony base.
Flesh: Thin and grey.
Habitat and season: Fairly common on leaf debris and in

grass, especially in damp places. Autumn.
Edibility: Worthless.

* **Leptonia incana** (Fries) Gillet POISONOUS
Rhodophyllus incanus (Fries) Kühner and Romagnesi

Cap: 1–3 cm. Umbilicato-convex then expanded. Striate, somewhat silky when dry. Olive or bronzy-green to grey-brown and zoned to a degree.
Gills: Distant and narrow, adnate or decurrent by a tooth. Greenish-white then flesh-colour.
Spores: Pink and angular. 10–13 × 6–9 μm.
Stipe: Short and very cartilaginous. Concolorous with cap or greenish-yellow. Hispid white at base. Bruises bluish.
Flesh: Green and thin. Has a strong odour of mouse droppings.
Habitat and season: Found in grass, open woodland and heaths. Fairly common from August to November.
Edibility: *Poisonous.*

* **Nolanea sericea** (Mérat) P. D. Orton
Entoloma sericeum (Mérat) Quélet
Rhodophyllus sericeus (Mérat) Quélat
Silky nolanea

Cap: 2–6 cm. Convex, soon expanded and flattish, often with a slight umbo. Margin usually splitting. Hygrophanous, grey to brown but drying much paler.
Gills: Distant and broadly emarginate, slightly adnexed. Equally attenuated from the stipe to margin. Greyish then rufescent.
Spores: Pink, angular and nodulose. 8–10 × 6–7 μm.
Stipe: Thin and shortish, hollow and shining. Often splitting lengthwise. Greyish.
Flesh: Umber-coloured. With a strong odour and taste of new meal.
Habitat and season: Usually gregarious or clustered in pastures, lawns, etc. Common from May to November.
Edibility: No.

* **Leptonia serrulata** (Fries) Kummer POISONOUS
Rhodophyllus serrulatus (Fries) Quélet
Eccilia atrides (Lasch.) Kummer

Cap: 2–3 cm. Convex then expanded. Fibrillose and blackish-blue at first but becoming brownish.
Gills: Adnate, very broad and crowded. Blue-grey then grey-flesh colour with a black serrulated edge.
Spores: Pink and angular. 10–12 × 6–7 μm.
Stipe: Slender and hollow. Pale cap-colour, dotted black at apex and white hispid at base.
Flesh: Whitish and thin.
Habitat and season: Grassy places and open woodlands from July to November.
Edibility: *Poisonous.*

* **Leptonia chalybaea** (Fries) Kummer POISONOUS
Rhodophyllus chalybaeus (Fries) Quélet

Cap: 2–3 cm. Steel-blue or dark violaceous.
Gills: Light bluish-grey with pale edge.
Spores: Pink. 12–13 × 6–7 μm.
Stipe: Dark blue.
Flesh: Deep blue.
Habitat and season: In grass. July to November.
Edibility: *Poisonous.*

* **Leptonia euchroa** (Fries) Kummer POISONOUS
Rhodophyllus euchrous (Fries) Quélet

Cap: Violaceous to dark purple.
Gills: Dark violet becoming paler except at edge.
Spores: 9–11 × 5–7·5 μm.
Habitat and season: May be found growing on stumps of frondose trees, especially alder and hazel. July to November.
Edibility: *Poisonous.*

* **Clitopilus prunulus** (Fries) Kummer
Sweetbread mushroom or The Miller

Cap: 3–8 cm. Convex then plane, finally depressed. Usually irregular and lobed, margin inrolled and mealy. Opaque-white, slimy in moist conditions.
Gills: Crowded, deeply decurrent, narrow and thin. Whitish then pale flesh-pink.
Spores: Pink and spindle-shaped with longitudinal ribs. 10–14 × 4·5–6 μm.
Stipe: Short, downy and striate, narrowing downwards or from midway downwards. White and sometimes eccentric.
Flesh: Soft and white. Odour and pleasant taste of new meal.
Habitat and season: Common in grassy places in woods and fields from July to November. When young it might be mistaken optically for *Clitocybe cerussata* or a similar species, but the odour of new meal is distinctive.
Edibility: Very good.

* **Hygrophoropsis aurantiaca** (Fries) Maire
Cantharellus aurantiacus [von Wulfen] Fries
Clitocybe aurantiaca ([von Wulfen] Fries) Studer
False chanterelle

Cap: 3–8 cm. Convex with involute margin then plane and finally depressed, regular or not. Dry and velvety. Bright orange-yellow or ochraceous-yellow, often whitish when growing in wet areas.
Gills: Rather crowded, thin and narrow, bifid and decurrent. Concolorous with cap.
Spores: White, ovate and smooth (dextrinoid). 5·5–7·5 × 3·5–5·5 μm.
Stipe: Cylindrical and smooth, often curved, sometimes

eccentric. Concolorous with cap.
Flesh: Rather tough and spongy, slightly paler than cap. Taste mild or slightly acrid.
Habitat and season: Very common on the ground in coniferous woods and on heaths. From August to November.
Edibility: Inferior quality.

Variety *pallida* Peck. Gills white then cream.

Genus-Paxillus

Medium to large species. The cap is fleshy and the margin is involute at first. Occasionally resupinate. The stipe is central, eccentric or wanting. The gills are decurrent and soft, they are often united near stipe (when present) and easily separable from flesh. Spores are brown and smooth.

Found growing on the ground or on wood. Worthless as food but taken by some.

199 **Paxillus atrotomentosus** (Fries) Fries

Cap: 5–25 cm. Convex to plane, margin involute. Sooty to reddish-brown. Dry and sometimes woolly.
Gills: Crowded, decurrent and narrow. Often uniting at base to form a network. Ochre-yellow.
Spores: Pale ochre and elliptical. 4·5–6·5 × 3·5–4·5 μm.
Stipe: Usually somewhat eccentric, short and ample. Covered with brown to black velvety down.
Flesh: Thick, firm and white. Has a pleasant odour but a bitter taste.
Habitat and season: Grows on or adjacent to pine stumps. Fairly common from August to November.
Edibility: No.

* **Paxillus panuoides** (Fries) Fries

Cap: 3–8 cm. Shell-shaped and bracket-like, irregular and lobed. Dingy yellow and may have a violaceous tinge. Non-stipitate and can be resupinate.
Gills: Crowded and narrow, decurrent to the base and anastomising. Yellowish in colour.
Spores: Brown. 4–5 × 3–3·5 μm.
Flesh: Thin with a fragrant odour.
Habitat and season: Found occasionally on decayed or worked soft-woods and sawdust heaps. From August to November or even later.
Edibility: No, but not poisonous.

200 **Paxillus involutus** (Fries) Karsten
Inrolled paxil or Brown roll-rim

Cap: 5–20 cm. Convex at first, then flattened and finally depressed. Shining rusty-yellow in colour. The margin is strongly inrolled and downy. In a moist atmosphere the cap becomes sticky and often has woodland debris firmly attached.
Gills: Crowded, narrow, branched and decurrent. Often cross connected near stipe. Yellowish-brown, bruising darker at the touch and staining the fingers.
Spores: Yellowish-brown. 8–10 × 5–6 μm.
Stipe: Short, solid and full. Cap-coloured. Centrally placed or slightly eccentric.
Flesh: Yellowish becoming darker when exposed to air. Tender and juicy, smelling fruity and acidulous.
Habitat and season: Common on the ground in all types of woodland especially in damp places. Grows from August to November.
Edibility: Pleasant taste but *should not be eaten raw* as uncooked specimens are poisonous to some.

201 Gomphidius glutinosus (Fries) Fries
Gomphus glutinosus (Fries) Kummer

Cap: 3–12 cm. Convex then plane or slightly depressed. Smooth and very glutinous, purple-fuscous or fuscous, often marbled with black.
Gills: Deeply decurrent, distant, forked and mucilaginous. Pale grey, later blackish-grey.
Spores: Brownish to olive, spindle-shaped and smooth. 18–22 × 5–7 μm.
Stipe: Whitish, slightly thickened and yellow at the base. Glutinous and fibrillose.
Flesh: White, but yellow at base of stipe.
Habitat and season: Common, growing in groups under conifers. From July to November.
Edibility: Good after washing off the gluten.

* **Gomphidius roseus** (Fries) Karsten

Smaller than *G. glutinosus* and with a deep rose-pink and slightly glutinous cap.

Gills: Whitish but darkening.
Spores: 17–21 × 5–5·5 μm.
Habitat and season: Common under pines from July to November.
Edibility: Yes.

* **Chroogomphus rutilus** (Fries) Miller
Gomphidius rutilus (Fries) Lundell
Gomphidius viscidus (Linnaeus) Fries

Cap: 5–8 cm. Dull brown with a Burgundy-coloured tint, slimy.
Gills: Olivaceous then purple.
Spores: 20–23 × 6–7 μm.
Habitat and season: Grows under pines from July to December.
Edibility: Yes.

Genus-*Boletus*

Soft, fleshy species which are, almost without exception, terrestrial. The hymenium borne in tubes, opening by pores on the underside of the cap (instead of gills). The flesh of the cap is easily separated from the tubes; a characteristic not shared with the Polypores which are mainly bracket-like and attached to wood. The stipe is centrally placed.

The Boletaceae are more closely allied to the Agaricaceae than to the Polyporaceae and some of them show colour changes when cut or bruised. The greater percentage are edible.

202 Boletus elegans Fries
Suillus grevillei (Klotzsch) Singer
Ixocomus flavus var. *elegans* (Fries) Quélet
Larch boletus

Cap: 5–15 cm. Hemispherical or campanulate, then expanded with a slight umbo. Very slimy. Golden-yellow, orange or orange-brown.
Pores: Small and round then angular. Sulphur-yellow then olive; bruising brown.
Tubes: Sulphur-yellow then olive. Short and adnato-decurrent.
Spores: Oblong and yellowish. 7–10 × 3–4 μm.
Stipe: With a cream-coloured ring which soon erodes. Tallish, equal and stuffed. Yellow but soon turns rusty-brown. The apex is either granular or faintly reticulated.
Flesh: Firm and yellow. Becoming soft and spongy.
Habitat and season: Common and gregarious. Found chiefly under larch from March to November.
Edibility: Good when young after removal of the cap cuticle.

203 Boletus parasiticus Fries
Parasitic boletus

Cap: 2–6 cm. Convex and downy. Often irregular and cracked. Brownish-yellow.
Pores: Large and angular. Pale lemon-yellow then olive-brown often with red patches.
Tubes: Short, adnate or slightly decurrent. lemon-yellow then olive-brown.
Spores: Said to be olive. Fusoid. 12–18 × 4–4·5 μm. Specimens rarely drop spores and exact spore colour is undetermined.
Stipe: Slender and granular, narrowing towards the base. Ochre-yellow, darker below with reddish streaks.
Flesh: Fibrous. Yellow in cap, reddish in stem.
Habitat and season: Parasitic on Scleroderma (earth-balls). Only occasionally found. September to November. The illustrated host is *Scleroderma citrinum* Persoon.
Edibility: Worthless. The taste is mild.

204 Boletus granulatus Fries
Suillus granulatus (Fries) O. Kuntze
Ixocomus granulatus (Fries) Quélet
Granulated boletus

Cap: 4–10 cm. Hemispherical or slightly conical then convex and finally plane. Slimy in wet weather. Reddish-brown or yellowish-brown, inclining more to yellow with age.
Pores: Yellow, small and angular. When young they exude milky drops which granulate as they dry.
Tubes: Yellow to greenish-yellow, short and adnate.
Spores: Narrowly oblong. Yellow-brown or clay colour. 9–10 × 3·5–4·5 μm.
Stipe: Pale yellow. The apex is covered with small yellow granules.
Flesh: Pale yellow, ample and tender. Colour unchanging on exposure to air.
Habitat and season: Very common and gregarious on the ground in pine woods. From August to November, occasionally earlier.
Edibility: Good when young after removal of the cap cuticle.

201 Gomphidius glutinosus

202 Boletus elegans

203 Boletus parasiticus

204 Boletus granulatus

205 Boletus edulis Fries
Cèpe, Edible boletus or Penny-bun bolete

Cap: 10–20 cm sometimes larger. Hemispherical then convex, often irregular in form. Glabrous, dry, shining or slightly viscid. Colour very variable, but basically brown through to white.
Tubes: Adnexed, long and thin. White then yellow or greenish. Easily detached from cap.
Pores: Small and round. Off-white then yellowish, finally greenish-yellow.
Spores: Medal-bronze and fusoid. 13–17 × 4–5·5 μm.
Stipe: Robust and solid, often swollen in the middle or below. Frequently irregularly shaped. Whitish or brown in colour, lighter at base, with fine white reticulations above, occasionally all over.
Flesh: Firm when young but soft in older specimens. White in the main, brownish immediately below cap cuticle. (Unchanging.)
Habitat and season: Common on the ground in, and at the perimeters of, frondose and pine woods. From August to November.
Edibility: Rated as one of the best edible species. It is mild and pleasant tasting and can also be dried or preserved in oil.

* **Boletus reticulatus** (Schaeff.) Boudier

Very similar to *B. edulis* but the cap is 10–15 cm and minutely granular or downy. The stipe is covered completely with a dense whitish network of raised lines. The spores are buffy-olive. Edible and excellent.

206 Boletus badius Fries
Xerocomus badius (Fries) E. J. Gilbert
Bay boletus

Cap: 5–12 cm. Bay-brown or darker. Hemispherical at first with margin involute. Later irregularly convex. Viscid in wet weather but velvety when dry.
Tubes and Pores: Adnate and variable in size but always angular. Whitish turning pale yellow later greenish-yellow. When bruised becoming greenish-blue.
Spores: Brownish-olive, fusoid. 13–15 × 4·5–6 μm.
Stipe: Longish, cylindrical and pale brown. Lighter at apex with white filaments at the base. Striate.
Flesh: Thick. White or pale lemon but turns azure blue.
Habitat and season: Grows mainly under conifers from August to November. Fairly common.
Edibility: Good with a mild taste.

* **Boletus calopus** Fries
Scarlet-stemmed boletus

Cap: 7–15(20) cm. Hemispherical then convex and somewhat gibbous. Dry and velvety at first, becoming glabrous. Whitish-grey to olive-brown.
Tubes: Adnate, thin and long. Lemon-yellow then paler, finally olive.
Pores: Minute and round. Yellow then olive-yellow, greenish-blue to touch.
Spores: Light olivaceous-brown, fusoid. 12–16 × 4·5–5·5 μm.
Stipe: Robust and solid. Scarlet with white or pinkish reticulations. Apex often yellow.
Flesh: Thick and whitish or pale yellow. Bluish when broken, especially in stipe. Odour pleasant. Taste bitter.
Habitat and season: Occasional, mainly under conifers. July to November.
Edibility: Worthless, extremely bitter.

207 Boletus erythropus (Fries) Secretan

Cap: 7–15 cm occasionally larger, hemispherical and then convex. Bay or dark brown minutely pubescent and dry. Sometimes with a reddish or olive tinge.
Tubes: Free, yellow then green.
Pores: Fine. Blood or orange red, bruising blue.
Spores: Medal-bronze, fusoid. 9–13 × 4–6 μm.
Stipe: Shortish and thicker at base. Brownish in colour but obscured by minute red dots, velvety, no net.
Flesh: Yellow but turns deep blue very quickly.
Habitat and season: Fairly common in all types of woodland from July to November.
Edibility: Good with a mild taste.

* **Boletus luridus** Fries
Lurid boletus

Very similar to *B. erythropus* but the stipe has a network of red and orange red. Not punctate.
Spores: Medal-bronze, fusoid. 12–15 × 5·5–6 μm.
Habitat and season: Found occasionally in frondose woods from August to November.
Edibility: Has a mild taste.

208 Boletus luteus Fries
Suillus luteus Fries
Yellow-brown boletus or Slippery jack

Cap: 4–10 cm. Conical, convex and then expanded. Very viscid or glutinous. Dark purplish-brown, rarely yellow. Radially streaked with fibrous lines. Pellicle easily separable.
Tubes and Pores: Short, adnate and angular, covered at first by a white membranous veil. Yellow and unchangeable, finally yellowish-olive.
Spores: Clay colour and sub-fusoid. 7–10 × 3–3·5 μm.
Stipe: Medium to slender with ring. Yellow and granular above ring. White or fuscous below. The ring is lax and white on the upper side, it often erodes with age.
Flesh: White or slightly yellow but pinkish at base of stipe. Unchangeable.
Habitat and season: Common under conifers from August to November.
Edibility: Good having a mild taste.

205 Boletus edulis

206 Boletus badius

207 Boletus erythropus

208 Boletus luteus

209 Boletus porosporus (Imler) Watling
Xerocomus porosporus Imler

Cap: 3·5–7·5 cm. Convex to plane. Dark olivaceous-brown with paler margin, then more sepia to cigar-brown. At first with yellowish tomentum which bruises when handled and darkens particularly at the margin. Finally cracking and at centre showing yellowish flesh.
Tubes: Adnate by a tooth. Lemon-yellow, finally flushed olivaceous. Bruising blue.
Pores: Compound, angular, lemon-yellow or dark sulphur-yellow, becoming darker with age. Bruising blue.
Spores: Olivaceous snuff-brown in the mass, elliptical. 13–15 × 4·5–5·5 μm. Usually exhibiting a distinct pore or truncation.
Stipe: Equal or attenuated upwards from the base. Apex lemon-yellow becoming flushed lemon-chrome with bay or brown-vinaceous to blood-red zone or not. Fibrillose-streaky below, slightly ribbed with olivaceous-brown and covered by olivaceous-brown to drab flecks, darkening on handling.
Flesh: Pale lemon-yellow to dirty buff with faint brown line under cuticle. Lemon-chrome in stipe apex. Dark brick colour or flushed brown-vinaceous at base, finally sepia, becoming blue. Taste and odour indistinctive.
Habitat and season: Widespread and probably not uncommon in frondose woods. Similar to *B. subtomentosus* and *B. chrysenteron* but differs in the sepia coloration when mature, greyish tinge to the pores and having a distinct pore in the spores which is unique in European Boletes. It is also found in N. America where it is less common and often confused with *Boletus truncatus* Snell & Dick.
Edibility: Not known.

210 Boletus mirabilis Murrill

Cap: 5–20 cm across. Convex to plane, densely woolly or hispid, dry. Very dark bay to dark greyish-brown.
Tubes: Long and yellow.
Pores: Angular and small. Depressed just at stipe. Bright sulphur-yellow, ageing to mustard-yellow.
Spores: Olive brown, smooth and elliptical. 14–24 × 6·5–9 μm.
Stipe: Rather long, slightly attenuated upwards, reticulated and bulbous. The white reticulation darkens with age and the stipe becomes more or less concolorous with cap, but whitish at base.
Flesh: Thick. White to yellowish.
Habitat and season: Common on or near decaying hemlock logs and stumps. August to November. Not recorded in Britain. Occurs in the Pacific N.W. of America. Rare in the Great Lakes area.
Edibility: Excellent.

211 Boletus versipellis Fries & Hök
Leccinum versipellis (Fries & Hök) Snell

Cap: 4–16 cm (We have found them up to 30 cm on occasion.) Hemispherical then convex. Tomentose when dry. Yellowish-orange to brick-red.
Pores: Dingy white then grey. Round and minute.
Tubes: Dingy white and long. Becoming very soft and anastomising.
Spores: Snuff-brown, smooth and spindle-shaped. 13–16 × 4–5 μm.
Stipe: Usually long, but sturdy and tapering upwards. Whitish or grey covered with blackish or reddish vertical lines or scales.
Flesh: White then pink. Becoming bluish-green at base of stipe. Smell and taste pleasant.
Habitat and season: Common on the ground in frondose woods especially birch. July to November.
Edibility: Good.

* **Boletus scaber** Fries
Leccinum scabrum (Fries) S. F. Gray

Very similar in form and shape to *B. versipellis*, but with cap light brown or grey to blackish.
Spores: Snuff-brown. 15–18 × 5–6 μm.
Edibility: Good.

* **Boletus aurantiacus** [Bull.] St Amans
Leccinum aurantiacum ([Bull.] St Amans) S. F. Gray

Again similar to *B. versipellis* but cut flesh turns slowly from grey to black.

N.B. There are many closely allied species that require close examination before identification can be determined and which we consider outside the scope of this book.

* **Boletus subtomentosus** Fries
Xerocomus subtomentosus (Fries) Quélet
Yellow-crack boletus

Cap: 4–10(15) cm. Hemispherical then convex to plane, margin undulate. Surface minutely felty or downy, usually cracking, especially at disc. Deep olive-brown or smoky-brown to greyish-fawn. The yellow flesh showing through the cracks.
Tubes: Adnate. Sulphur-yellow then olive-yellow.
Pores: Large, unequal, denticulate and angular. Bright golden-yellow and remaining so but sometimes bruising slightly blue.
Spores: Medal-bronze and fusoid. 11–15 × 4·5–5 μm.
Stipe: Slender and often attenuated at base. Yellowish with slightly raised reddish-brown ribs.
Flesh: Firm and pale yellow, sometimes changing faintly blue when broken. Taste pleasant, odour mild.

209 Boletus porosporus

210 Boletus mirabilis

211 Boletus versipellis

212 Boletinus cavipes

Habitat and season: Common in both coniferous and frondose woods, usually in grassy places. From August to November.
Edibility: Good.

* **Boletus chrysenteron** St. Amans
Xerocomus chrysenteron (St. Amans) Quélet

Cap: 4–10 cm. Hemispherical then convex to plane, minutely tomentose. Fawn to reddish-brown or with olivaceous tint. Often cracking to expose purplish-red flesh under cuticle.
Tubes: Adnate. Sulphur-yellow then dirty olive.
Pores: Large, angular and compound. Yellow then olivaceous, bruising slowly bluish-green.
Spores: Medal-bronze and fusoid. 12–15 × 4·5–5 μm.
Stipe: Rather slender and often attenuated at base and apex, curved. Fibrous and more or less furrowed yellowish, strongly streaked with purplish-red in the upper part.
Flesh: Creamy yellow. Purplish-red under cap cuticle and often also in the stipe. The yellow flesh bluing slightly when broken. Has a mild taste and a pleasant odour.
Habitat and season: Common in all types of woodland from August to November.
Edibility: Fair, but becomes pulpy on cooking.

* **Boletus sulphureus** Fries
Phlebopus sulphureus (Fries) Singer

Cap: 7–12 cm. Convex then plane, silky tomentose, usually dry. Sulphur-yellow.

Tubes: Adnate. Greenish-yellow. Rusty spotted with age.
Pores: Minute, compound. Sulphur-yellow, then greenish.
Spores: Medal-bronze and elliptical. 6–7 × 3 μm.
Stipe: Thick, firm and ventricose. Striate and wrinkled. Sulphur-yellow then ferruginous.
Flesh: Sulphur-yellow, turns greenish-blue when broken and later golden yellow. Flesh in stipe base often red.
Habitat and season: Grows in clusters under pines or on sawdust and wood chippings etc., from a golden yellow mycelium. Rare. Found from July to October.
Edibility: Yes.

* **Boletus aeruginascens** Secretan
Boletus viscidus Fries & Hök
Suillus aeruginascens (Secretan) Snell
Viscid boletus

Cap: 6–10 cm. Hemispherical or campanulate then convex to plane. Very viscid and slightly rugose. Livid white, dirty yellowish or greenish-ashy. In dry weather it becomes scaly. Cuticle easily detached.
Tubes: Sub-decurrent and very long. Greyish to violaceous-grey.
Pores: Large, compound, unequal, angular and dentate. Off-white then violaceous-grey and finally brown.
Spores: Snuff-brown and sub-fusoid. 10–13 × 4·5 μm.
Stipe: Equal and viscid. White then yellowish or grey reticulate above the ring.
Ring: Ample, thin, membranous and white. Soon eroding leaving a brownish zone on stipe.
Flesh: White in cap, greenish or yellowish in stipe, bluing slightly when broken. Taste and odour agreeable.
Habitat and season: Fairly common under larches from June to November.
Edibility: Mediocre; always discard cap cuticle before cooking.

* **Boletus bovinus** (Linn.) Fries
Ixocomus bovinus (Fries) Quélet
Suillus bovinus (Fries) O. Kuntze

Cap: 4–8 cm. Convex then plane. Usually irregular and cracking. Viscid and smooth. Reddish-buff, margin white when viewed from below.
Tubes: Decurrent. Yellow to rusty, finally olive-brown.
Pores: Large, compound and angular.
Spores: Light brownish-olive, sub-fusoid. 8–10 × 3–4·5 μm.
Stipe: Slender and smooth, narrowing downwards. Pale brown.
Flesh: Yellow to pinkish. Mild taste and pleasant odour.
Habitat and season: Common under pines, often with *Gomphidius roseus* (Fries) Karsten. From July to November.
Edibility: Good when young, but soon infected by insect larvae.

* **Boletus piperatus** Fries
Ixocomus piperatus (Fries) Quélet
Suillus piperatus (Fries) O. Kuntze

Cap: 2–6 cm. Convex then plane. Smooth and slightly viscid. Cinnamon.
Tubes: Adnate or sub-decurrent. Yellow to rusty.
Pores: Wide and angular. Coppery-red to brown.
Spores: Snuff-brown and elliptical. 8–11 × 3–4 μm.
Stipe: Slender. Cinnamon with yellow fusiform base. Mycelium yellow.
Flesh: Chrome yellow, rhubarb-coloured in cap. Odour indistinctive.
Habitat and season: Common in mixed woods and on heaths. August to November.
Edibility: Acrid and worthless.

* **Boletus impolitus** Fries
Xerocomus impolitus (Fries) Quélet

In appearance rather like a large *Boletus subtomentosus*.

Cap: 6–20 cm. Hemispherical then convex, finally plane. Floccose finally rivulose. Pale clay-colour to tawny-olive, browning when handled.
Tubes: Adnate or free. Sulphur-yellow.
Pores: Minute. Sulphur-yellow, greenish to touch.
Spores: Medal-bronze, fusoid. 10–14 × 4·5–5·5 μm.
Stipe: Robust. Even paler than cap, often reddish-yellow near apex. Not reticulated.
Flesh: Thick and white or tinged yellowish, unchanging. Mild taste. Odour of vinegar, unpleasant.
Habitat and season: Occasional on base rich clayey land. Under frondose trees, especially oak.
Edibility: Yes.

* **Boletus variegatus** Sow. ex Fries
Ixocomus variegatus (Sow. ex Fries) Quélet
Suillus variegatus (Sow. ex Fries) O. Kuntze
Variegated boletus

Cap: 6–12(15) cm. Hemispherical then convex to plane and often irregular. Slimy in moist weather. Ochraceous with small soft darker scales.
Tubes: Adnate. Pallid then yellowish or cinnamon.
Pores: Rounded. Small to medium in size. Cinnamon, bruising light blue.
Spores: Medal-bronze, sub-fusoid. 9–11 × 3–4 μm.
Stipe: Robust and solid. Paler than cap and with a reddish-brown base.
Flesh: Pale yellow, reddish at base of stipe. Turns blue in parts on breaking. Taste mild. Odour unpleasant.
Habitat and season: Common and gregarious under conifers. August to November. Not recorded in America, but closely related to *Boletus tomentosus* Kauffman.
Edibility: Inferior but not poisonous.

* **Boletus satanas** Lenz POISONOUS
Satan's boletus

Cap: 10–20 cm. Hemispherical, then convex to plane. Polished and smooth when mature. Whitish-grey or pale olive.
Tubes: Free. Yellow then green.
Pores: Small. Yellow at first, becoming blood-red then orange-red and finally brown.
Spores: Medal-bronze, fusoid. 11–13 × 4–5 μm.
Stipe: Robust, short and squat. Yellow to reddish with blood-red reticulations.
Flesh: White to cream, changing slightly blue when broken. Taste mild. Odour unpleasant.
Habitat and season: Occasional under frondose trees. July to November.
Edibility: *Poisonous.*

* **Boletus pulverulentus** Opat.
Xerocomus pulverulentus (Opat.) Moser apud Gams.

Cap: 4–10(14) cm. Hemispherical, then expanded. Dry, often lobed, minutely tomentose. Yellow at first but soon becoming reddish-brown and finally black.
Tubes: Adnate. Lemon-yellow to gold.
Pores: Rather large. Yellow, bruising blackish-blue.
Spores: Medal-bronze, fusoid. 11–14 × 3·5–5 μm.
Stipe: Robust, fusiform. Slightly velvety. Orange-yellow above, reddish-brown below, black to touch.
Flesh: Yellow in cap and upper stipe, turning blackish-blue when broken. Flesh reddish in stipe base. Taste and odour agreeable.
Habitat and season: Occasional in damp places under frondose trees, especially on calcareous substrate. August to November. It is rarely observed in its brighter colours as all parts soon turn black.
Edibility: Good.

* **Boletus cramesinus** Secretan
Boletus tenuipes (Cooke) Massee

Cap: 2–5 cm. Hemispherical, then convex to plane. Viscid and slightly rugose. Pink or 'crushed strawberry' in colour.
Tubes: Adnate or slightly decurrent. Chrome-yellow.
Pores: Fairly large and angular. Vivid gold.
Spores: Clay-colour, fusoid. 11–15 × 4·5–5·5 μm.
Stipe: Slender, attenuated at base. Striato-fibrillose.
Flesh: Whitish to pale pink. Odour and taste agreeable.
Habitat and season: Occasional under frondose trees, especially on burnt ground. July to November. Not recorded in N. America, but closely related to *Boletus auriporus* Peck.
Edibility: Fair to good.

212 **Boletinus cavipes** (Opatowski) Kalchbr.
Suillus cavipes (Opatowski) Smith & Thiers
Boletus cavipes Opatowski
Hollow-stemmed boletus or Mock oyster

Cap: 3–14 cm. Onion-shaped at first with an acute umbo, then convex, finally expanded, uneven and centrally depressed. Felty scaly, young specimens often having viscid white velar remains hanging from margin. Individual coloration very variable, from lemon colour to dark or violaceous-brown.
Tubes: Short and very decurrent. Pale yellow becoming greenish.
Pores: Large, angular and honeycomb-like. Pale yellow to greenish-yellow.
Spores: Greenish-yellow and sub-fusoid.
Stipe: Fragile and always hollow. Often swollen midway, curved at base and rooting. Slightly lighter in colour than cap. Reticulate above the ring and scaly below it.
Ring: White, thick and viscid. Floccose and evanescent.
Flesh: Soft and yellow in cap. Fibrous and with rosy shades in stipe. Mild taste and pleasant odour. Unchanging.
Habitat and season: Singly and gregarious in mixed woods, but especially larch. August to November. Rare in the British Isles.
Edibility: Fairly good when young. It can be preserved and dried.

213 Suillus subolivaceus Smith and Thiers

Cap: 5–10(15) cm. Convex then plane or obtuse, expanding to broadly umbonate. Very glutinous and appears streaked beneath by agglutinated fibrils. Snuff-brown to olive-brown or dingy olive.
Tubes: Adnate becoming more or less decurrent. Dingy yellow or olive-greyish.
Pores: Concolorous or slightly browner. Unchanging to touch.
Spores: Dingy cinnamon in the mass, subfusoid and smooth (8)9–11 × 3–4(4·5) μm.
Stipe: 6–10 cm long, 8–14 mm thick. Solid and equal. Yellowish above, paler below, covered with pinkish-brown dots but blackens through handling.
Ring: Median to superior, shrinking quickly after breaking. Membranous with outer gelatinous layer. Concolorous with cap.
Flesh: Thick and spongy. Pallid to yellowish or olivaceous-grey, no colour change when broken. Odour mild, taste acidulous.
Habitat and season: Gregarious under conifers in parts of N. America during summer and autumn, easily confused with *S. subluteus* (Peck) Snell ex Slipp and Snell. Not recorded in Britain.
Edibility: Good.

214 Suillus sibiricus (Singer) Singer
Boletus sibiricus (Singer) Smith
Ixocomus sibiricus Singer

Cap: 3–10 cm. Convex then plane or slightly umbonate, viscid. Ground colour chamois to dingy olive-yellow beset with adhering snuff or cinnamon-brown adpressed scales (veil tissue), especially at the margin where they form a soft pale yellow cotton roll, which disappears with age.
Tubes and Pores: Adnate becoming decurrent. Dingy ochraceous to almost honey-yellow, bruising vinaceous-cinnamon. Mouths angular and often compound.
Spores: Dull cinnamon, narrowly elliptic in face view. 8–11 × 4–4·5 μm.
Stipe: 5–10 cm, long and 7–15 mm thick at apex. Dingy ochre-yellow above, soon stained vinaceous at the base. Glandular dotted overall. Sometimes with a ring.
Flesh: Pale olive-yellow. Slowly becoming dull cinnamon when cut. Taste acidulous.
Habitat and season: Gregarious under white pine in the Pacific N.W. of America. Not recorded in Britain.
Edibility: Good.

* **Suillus americanus** (Peck) Snell ex Slipp & Snell
and
* **Suillus umbonatus** Dick & Snell

Both these species are very similar to *S. sibiricus* and advanced literature should be referred to for detailed differences.

215 Strobilomyces floccopus (Fries) Karsten
Strobilomyces strobilaceus (Fries) Berk.
Old man of the woods or Cone-like boletus

Cap: 7–15 cm. At first almost spherical but later convex. Dry with broken surface of pointed overlapping scales which are brown or blackish. Grey interspaces. The cap margin ends in a flocculose fringe.
Pores: White, large and polygonal and covered at first by a membranous veil; finally blackish. Young specimens turn red on bruising.
Tubes: Long but shorter near to stipe. They are whitish-grey, adnate or free.
Spores: Black-purple, sub-globose. 10–13 × 8–10 μm, reticulate.
Stipe: May be long or short. White above ring, scaly below.
Ring: Sheathed and white but soon disappears.
Flesh: Tough and white at first. Later reddish and finally blackening with age.
Habitat and season: Woods generally but not common. August to November.
Edibility: No, worthless as an esculent.

Genus-Russula

Includes small, medium and large species which are often brightly coloured.

The cap is convex, flat or depressed. The gills are adnate, rigid and fragile; the spores are white to deep ochre-yellow, with amyloid ornamentations. The stipe is centrally positioned and fleshy. The flesh of the stipe is continuous with that of the cap, brittle and granular.

They are chiefly terrestrial species growing near to or under trees. The mild tasting species are edible.

In some instances identification is possible by the application of ferrous sulphate to the flesh when a colour change is apparent.

216 Russula queletii Fries apud Quélet

Cap: 4–8 cm. Convex, striate and viscid, then plane and dry. Wine-red shading to purple-lilac at margin.
Gills: Adnate. White at first but finally greyish. Alternated, unequal or forked. Exuding droplets which on drying form azure-blue or pale olive spots.
Spores: Cream-ochre. 7·5–9 × 6·5–7·5 μm.
Stipe: Wine-red, spongy and mealy. Usually shorter than width of cap.
Flesh: White but reddish under pellicle. Thin and firm. (No reaction when treated with an alkaline solution.)
Habitat and season: Locally common under conifers or frondose trees from August to November.
Edibility: No, very acrid.

213 Suillus subolivaceus

214 Suillus sibiricus

215 Strobilomyces floccopus

216 Russula queletii

* **Russula sardonia** similar to *R. queletii* but found under conifers. (Turns bright red when treated with a 50% solution of Ammonia in distilled water.)

* **Russula nitida** (Fries) Fries

Cap: 3–5 cm. Purplish-brown.
Gills: Pale and crowded, soon becoming bright yellow.
Spores: Ochraceous. 9–10 × 7–8 μm.
Edibility: No.

217 Russula nigricans (Mérat) Fries
Black russule

Cap: 8–20 cm. Convex, then depressed. Dirty white at first becoming smoky brown and finally more or less black. Firm, stout and coarse.
Gills: Adnate, distant and thick with intermediates. White at first becoming reddish-brown at edges.
Spores: White. 7–10 × 6·5–9 μm.
Stipe: Cylindrical and squat. White at first later brownish to black.
Flesh: Thick, hard and coarse, changing colour to reddish (see illustration) and finally blackish when exposed to the atmosphere. With age the flesh in both cap and stipe becomes brownish-black and the entire fruit-body assumes a blackish cast remaining in this mummified state until the first heavy rain causes decay.
Habitat and season: Found predominantly in frondose woods where it usually occurs scattered over a small area. Common from July to November.
Edibility: Of a most inferior quality.

* **Russula albonigra** (Krombh.) Fries

Similar to *R. nigricans* but the gills are crowded and unequal. In young specimens the flesh turns greyish on exposure to air. The whole fungus bruises black to the touch.

* **Russula adusta** (Fries) Fries
The Scorched russule

This also has crowded gills but the flesh does not redden on exposure to air, but simply becomes brownish.

* **Russula densifolia** (Secretan) Gillet

Similar in appearance and behaviour to *R. nigricans* but like *R. albonigra* in having crowded gills.

218 Russula brevipes Peck

Cap: 9–20 cm. Disc depressed at first and with incurved margin, then expanding to broadly infundibuliform and very often soiled by debris. Dry and minutely tomentose. At times somewhat radially wrinkled, margin not striate. White to buffy-white, finally spotted or stained rusty-brown, or even brownish overall.
Gills: Narrow, thin acute in front, decurrent with intermediates. Moderately distant to occasionally crowded, connected by veins. Nearly white at first, then cartridge-buff, sometimes with a greenish tinge.
Spores: White to light cream in mass. Usually broadly elliptic, occasionally broadly ovate or sub-globose. 8–10·5 × 6·5–9·5 μm but sometimes larger.
Stipe: 3–8 cm long. Equal or narrowing towards base, straight and unpolished. Solid, then hollowed by insect larvae to which it is prone. White then stained brown.
Flesh: Firm, white and thick. Mild tasting. Stained brown where attacked by insect larvae.
Habitat and season: A common North American species found in frondose or pine woods from July to November. The cap usually grows level with, or only slightly above the soil surface; therefore, the stipe is seldom if ever visible. Not recorded in Britain.
Edibility: Of poor quality.

* **Russula delica** Fries

Is almost identical to *R. brevipes* therefore needs no further description.
Spores: 8·5–9 × 6·5–7 μm.
It is a common British species and might be passed over as *Lactarius vellereus* (Fries) Fries, whose flesh exudes white milk when broken.

219 Russula decolorans (Fries) Fries

Cap: 5–15 cm. Spherical then expanded and depressed, very regular. Viscid in moist weather. Orange-red at first then browny-orange and finally blackish. Margin becoming striate.
Gills: Slightly adnexed, often in pairs. Crowded, thin and fragile. White at first, then greyish or creamy.
Spores: White tinged ochraceous, sub-globose with long spines. 10–14 × 9–12 μm.
Stipe: Tall and stout, usually rugosely striate. White then grey, especially inside.
Flesh: Rather thick. White then grey, finally black-spotted. Odour and taste mild.
Habitat and season: Under pines or in mixed woodlands, especially wet areas. Grows singly or gregarious. Occasional from August to November.
Edibility: Yes.

220 Russula foetens (Fries) Fries
Stinking russule

Cap: 6–12 cm. Sub-globose, convex, then plane or depressed. Pellicle not separable. Viscid in damp weather. Sulcate for a considerable distance from the margin, which is at first involute. Of a dingy ochraceous colour.
Gills: Adnexed or free. Crowded and forked with numerous intermediates connected by veins. When young exuding drops of water. Whitish, edged or spotted brownish with age.
Spores: Pale cream, broadly elliptical and covered with small spines. 9–10 × 8 μm.
Stipe: Thick and rigid, soon becoming hollow. Whitish becoming stained brownish.
Flesh: Thin, rigid, fragile and pallid. Taste acrid. Usually smells foetid but not always.
Habitat and season: Fairly common in mixed woodlands. August to November.
Edibility: Worthless.

217 Russula nigricans

218 Russula brevipes

219 Russula decolorans

220 Russula foetens

* **Russula laurocerasi** Melzer

Cap: 4–7 cm. Similar to *R. foetens* but smaller. Smells of bitter almonds.

* **Russula consobrina** (Fries) Fries
Russula livescens (Batsch) Quélet

Cap: 3–9 cm. Fuliginous brown and viscid with an even membranous margin.
Gills: Free, crowded and forked, with intermediates. Clear white.
Spores: Pale ochraceous, almost spherical, warted. 9–11 × 8–9 μm.
Stipe: Robust and equal. Clear white becoming grey.
Flesh: White with acrid taste. Thin at disc and grey below the thick separable cuticle.
Habitat and season: Coniferous woods, August to November.
Edibility: No.

221 Russula ochroleuca (Secretan) Fries
Yellow russule

Cap: 4–9 cm. Convex, then flat and finally depressed. Smooth, viscid in wet weather. Striate at margin. Yellowish-ochre but fading with age.
Gills: Adnexed, broad and rather distant. White at first, becoming creamy.
Spores: Creamy. 8–10 × 6·5–8 μm.
Stipe: White and robust but brittle, reticulated veined. Later spongy tinged with grey.
Flesh: White but yellowish under pellicle. Coarse in texture.
Habitat and season: A very common fungus in woodlands from August to November.
Edibility: No, has a sharp taste and granulose texture.

* **Russula lutea** Fries

Similar to *R. ochroleuca* but differs in having crowded gills which become yellow with age and are connected by veins. The flesh is mild to taste. Edible.

* **Russula claroflava** Grove

Cap 6–12 cm and of bright chrome-yellow. The gills are white to primrose and not crowded. Stipe bruises grey. Flesh white but slowly turns grey. An edible species usually found in wet ground under frondose trees.

* **Russula fellea** Fries

Cap 3–7 cm, straw coloured as is the entire fruit-body. The gills are adnate and crowded. Very acrid to taste and smells of geranium. Inedible. Found under frondose trees, especially beech, from August to November.

222 Russula cyanoxantha (Secretan) Fries
(illustration variety: *peltereaui*)

Cap: 5–12 cm. Globular at first becoming convex and finally depressed. Usually of a violet hue with variations of greens, greys and blues varying in depth. Sometimes of a rosy colour on the disc.
Gills: White, elastic and greasy to the touch. Adnexed, broad and thick, frequently branching and sometimes cross-connected by veins.
Spores: White. 7·5–9 × 6–7 μm.
Stipe: White and occasionally tinged with violet. Variable in length and thickness, becoming grooved with age. Elastic and solid, then spongy and hollow. Quite often eaten by insect larvae.
Flesh: White but pinkish under cap cuticle. Solid but somewhat granular in texture.
Habitat and season: Fairly common in frondose and coniferous woodlands from July to November.
Edibility: Has a sharp taste when eaten raw. Good when grilled.

223 Russula emetica (Fries) S. F. Gray
The Sickener

Cap: 5–10 cm. Hemispherical, convex, then plane or depressed. Glutinous when young. Scarlet when in prime condition but later becoming pink and may even degenerate to yellow or white. The margin becomes striate and the cuticle easily separable from the cap revealing superficially red flesh.
Gills: Broad, equal and rigid. Adnate or free. Crowded or distant. Shining white.
Spores: White, almost spherical and coarsely warted. 8–10 × 8–9 μm.
Stipe: Usually pure white but occasionally with a pinkish flush. Sturdy but very brittle, as is the whole fruit body.
Flesh: Reddish under cap cuticle, otherwise white.
Habitat and season: Common and gregarious, on the ground, especially in damp mossy places, mainly under conifers. From August to November.
Edibility: No. It has a very acrid and lingering taste, causes vomiting when eaten raw, and although it is sometimes used as hot seasoning with other fungi, we strongly advise against any such experiments.

* **Russula puellaris** Fries

Cap: 3–5 cm. Convex but very soon plane and slightly depressed at the centre. At first dark purple but soon becoming paler or yellowing.
Gills: Adnate, but very much narrowed behind, subdistant. White but soon yellowish.
Spores: Yellowish, sub-globose and echinulate. 9–10 × 7–8 μm.
Stipe: Short, soft and fragile. Rugose under a lens. White becoming yellowish. Soon hollow.
Flesh: White at first but turning yellow. Mild taste and odour.
Habitat and season: Common in mixed woodland usually in troops. August to November.
Edibility: Worthless.

* **Russula versicolor** J. Schaeffer

Similar in size and habitat to *R. puellaris* but cap usually greenish at the centre and the flesh acrid to taste.
Spores: Ochraceous, broadly elliptical, finely warted or with imperfect network. 8–9 × 5·5–6 μm.
Edibility: Worthless.

224 Russula xerampelina (Secretan) Fries

Cap: 5–12 cm. Convex, then flattened and finally depressed. Varying in colour from purple-red to brown. Viscid or dry.
Gills: Adnexed and rather crowded, broader in front and forked behind. White to yellowish.
Spores: Ochre, 8–9·5 × 7–8 μm.

221 Russula ochroleuca

222 Russula cyanoxantha

223 Russula emetica

224 Russula xerampelina

Stipe: Stout and firm eventually becoming spongy and hollow. White or reddish in colour, reticulately veined.
Flesh: Yellowish-white and compact.
Habitat and season: Found commonly in woods from July to November.
Edibility: Yes.

This is a variable fungus but it is readily identified by its shrimp or crab odour when adult. Treated with ferrous sulphate crystals or solution the flesh changes colour to a deep olive-green.

225 Russula mairei Singer
Maire's russule

Cap: 3–7 cm. Scarlet cap becoming creamy with age. Similar to but smaller than *R. emetica*. Cuticle half-peeling showing white flesh.
Gills: White and adnate.
Spores: White. 7–8 × 5·5–6 μm.
Stipe: White and sturdy. No longer than width of cap.
Flesh: White with a faint odour of honey.
Habitat and season: Common in woods especially beech. From August to November.
Edibility: Acrid and not recommended.

226 Russula atropurpurea (Krombholz) Britz.
Purple russule

Cap: 5–10 cm. Convex, then flat and finally depressed. Dark reddish-purple inclining to black at the centre. Slimy when moist. Fading with age and showing a yellow discolouration.
Gills: Adnexed or slightly decurrent. Broad and white but becoming greyish often with rusty coloured spots.
Spores: Creamy-white. 8–9 × 7–7·5 μm.
Stipe: Shortish and stout. White but often rusty at the base, becoming grey and spongy with age.
Flesh: White and thick.
Habitat and season: All types of woodland but especially in frondose woods. A very common species appearing from late July to November.
Edibility: Of coarse texture and slightly acid taste. Not recommended.

227 Russula sanguinea (St. Amans) Fries
Russula rosacea (Secretan) Fries

Cap: 3–10 cm. Convex then plane or slightly centrally depressed, margin even. Cuticle not separable. Viscid only in moist weather. Colours variable: blood-red through to pink or even white with age.
Gills: Crowded, narrow and usually slightly decurrent. White to cream and eventually spotted ochraceous-yellow.
Spores: Creamy. 7–9 × 6–7·5 μm.
Stipe: Tall and thick. Wrinkled and striate. Reddish, only rarely white.
Flesh: White and cheesy but reddish under cap cuticle. Has an acrid taste but a fruity odour.
Habitat and season: Occasional under conifers from August to November.
Edibility: No. Bitter and acrid.

* **Russula virescens** Fries
Green russule

Cap: 5–12 cm. Hemispherical, soon expanded and plane or even depressed. Often uneven and cracking exposing white flesh beneath the cuticle which is flocculose or coarsely mealy. Colour deep or pale green and often broken up into patches.
Gills: Crowded and free or nearly so, thick and forked near the stipe. White then creamy, brown-spotted with age.
Spores: Yellowish-white. 6·5–9 × 5–6·5 μm.
Stipe: White, short and stout. Slightly attenuated downwards.
Flesh: White and firm, then cheesy with age. Odour agreeable.
Habitat and season: Fairly common and gregarious on the ground in mixed woods, but mainly under frondose trees.
Edibility: Good, cooked or raw.

* **Russula heterophylla** (Fries) Fries
Russula furcata (Lambark ex Fries) Fries
Fork-gilled russule

Cap: 5–12 cm. Convex then plane, finally depressed. Viscid or dry, polished, or rugose and velvety except at disc. Often cracking. Margin striate only when aged. Bright green, olive-green or yellowish-green, occasionally with a tinge of grey or lilac. Becoming paler in the centre.
Gills: White, crowded, decurrent and very narrow, especially near stipe where they join together irregularly and on which they are decurrent by fine veins.
Spores: White in the mass, broadly elliptical and warted. 7 × 5–5·5 μm.
Stipe: White, stout, solid, firm, finely striate and more or less equal.
Flesh: White and firm. Taste mild.
Habitat and season: On the ground in frondose or coniferous woodland. From August to November.
Edibility: Good.

Genus-*Lactarius*

Small, medium and large species occur in this genus; they are often showy in appearance. The cap is convex, flat or depressed, and fleshy. The gills are decurrent and often crowded. The spores are white to deep cream with amyloid ornamentation; the stipe is central. The flesh is brittle and granular, yielding milk when broken. The taste and odour are important in identification. Mainly terrestrial growing near or under trees. The mild tasting species are edible.

228 Lactarius deliciosus (Fries) S. F. Gray
Delicious lactarid or Saffron milk-cap

Cap: 4–12 cm across. Convex but umbilicate and later depressed. Can be viscid or dry and has an inrolled and sometimes fleecy margin. Colour reddish-orange or orange-grey, with concentrical zones of a deeper hue. Becomes greenish with age.
Gills: Decurrent or adnate also narrow, crowded and

225 Russula mairei

226 Russula atropurpurea

227 Russula sanguinea

228 Lactarius deliciosus

forked. At first they are carrot-coloured becoming blotched or spotted with green.
Spores: Pinkish-buff. 7–9 × 6–7 μm.
Stipe: Short, central and cylindrical, stuffed or hollow. Often pitted. Concolorous with cap and gills.
Flesh: Brittle, white then orange with green stains.
Milk: Orange-red, copious and mild to the taste, turning green.
Habitat and season: Found mainly under conifers from August to November.
Edibility: Overrated in our opinion; its name is not a true description. Young specimens only should be sought and these grilled quickly; alternatively they may be salted or pickled.

* **Lactarius rufus** (Fries) Fries
Rufous milk-cap

Cap: 4–8 cm. Convex but soon depressed, then infundibuliform, usually with papilla, silky then polished. Margin involute at first. Bay-brown rufous.
Gills: Broad, crowded and adnato-decurrent. Whitish then more or less cap-coloured.
Spores: White and broadly elliptical. 8–9·5 × 6–7 μm.
Stipe: Concolorous with cap. Longitudinally furrowed and slightly pruinose. Stuffed then hollow.
Flesh: Pallid.
Milk: White and fairly copious, very acrid. Unchanging.
Habitat and season: Very common on the ground under conifers. From August to November.
Edibility: No.

229 Lactarius turpis (Weinm.) Fries
Lactarius plumbeus (Fries) S. F. Gray
Sombre lactarid, Base toadstool or Ugly toadstool

Cap: Under a coat of olive mucilage it is umber in colour, becoming darker with age. The margin is tawny and downy. When young the margin is inrolled or turned down. Diameter usually 8–12 cm but we have found specimens up to 30 cm across.
Gills: Crowded, forked and decurrent, yellowish or greyish-white becoming darker with age.
Spores: Cream. 7·5–8 × 6–7 μm.
Stipe: Olive, paling to whitish at the top and slimy. Robust and very short in proportion to the cap, tapering from apex down.
Flesh: Dirty white and somewhat brittle, especially in older specimens. It changes colour to purple immediately when treated with Ammonium hydroxide.
Milk: White, copious and very acrid.
Habitat and season: Common from August to November in deciduous woodlands, especially under birch in damp acid soils.
Edibility: Best avoided as it is very acrid.

* **Lactarius pyrogalus** (Fries) Fries

Cap: 4–9 cm. Convexo-plane then deeply depressed or infundibuliform. Dull greyish-brown, often tinged violet. Slightly viscid in moist weather. Often zoned.
Gills: Decurrent, narrow and sub-distant. Pale yellow to deep ochre.
Spores: Pale ochre and almost spherical. 6·5–8 × 5·5–6·5 μm.
Stipe: Equal and often attenuated downwards. Pale cap-coloured.
Flesh: White but greyish under cap cuticle.
Milk: White and very acrid. Turns orange-yellow with a 10% solution of Potassium hydroxide.
Habitat and season: Frequently found in grass near hedgerows or in woods, especially under hazel. September to November.
Edibility: No.

230 Lactarius fulvissimus Romagnesi
Lactarius ichoratus (Secretan) Fries

Cap: 4–7 cm. Obtuse, expanded then depressed. Rigid then soft. Tawny brick-red. Disc often brown.
Gills: Adnate with a decurrent tooth. Scarcely crowded, and narrow. Whitish then ochraceous, never spotted.
Spores: Deep cream. 8–10 × 6–7 μm.
Stipe: Equal or tapering downwards. Smooth and more or less concolorous with the cap.
Flesh: Pallid.
Milk: Mild and fairly copious. White and unchanging.
Habitat and season: Occasional in frondose woods from August to October.
Edibility: Not known.

* **Lactarius quietus** (Fries) Fries
Mild mushroom or Oak milk-cap

Cap: 4–9 cm. Convex, finally expanded and depressed. Viscid at first but later becoming silky. Cinnamon or dull reddish-brown in colour, the centre is darker than its surroundings which are concentrically zoned.
Gills: Slightly decurrent. Whitish at first but later becoming more or less colour of cap.
Spores: Pale pink. 7·5–9 × 6–7 μm.
Stipe: Rubiginous and stuffed. Darker at the base and longitudinally furrowed. Usually longer than width of the cap.
Flesh: Reddish-brown and fairly thick. Has a rancid oily smell.
Milk: White and mild with a pleasant sweet taste which is said to be reminiscent of walnuts.
Habitat and season: Common under oaks from September to November.
Edibility: Not recommended but is eaten by some.

231 Lactarius scrobiculatus (Fries) Fries

Cap: 7–18 cm. Convex but centrally depressed and finally infundibuliform. Very viscid when moist, with glutinous down. Yellow or light brownish-yellow in colour, but fading when in sunny situations. Usually without zones, but it is stated that it may be conspicuously zoned. The margin is incurved and fringed with shaggy fibrils but these are lost with age.
Gills: Slightly decurrent, thin and crowded, with intermediates. Whitish with yellow edges at first. Darker and spotted with reddish-brown when old.
Spores: Cream in the mass. Sub-globose and minutely echinulate. 7–9 × 6–7·5 μm. Amyloid.
Stipe: Stout and solid at first, but soon becoming hollow. Pitted with ochraceous erosions. Colour paler than the cap and somewhat viscid with a downy base.
Flesh: Thick, hard and fragile. White but yellowish at base of stipe. Acrid taste.
Milk: Copious and white, but quickly becoming yellow when exposed to the air.
Habitat and season: Locally common in damp woodlands. September to November.
Edibility: No.

232 Lactarius lignyotus Fries

Cap: 2–7(10) cm. At first conical to umbonato-convex, then expanded, finally depressed around the small umbo, margin involute. Dry and velvety, blackish or blackish-brown, rarely olive or dingy yellow-brown when faded. Young specimens have a narrow whitish band at the margin.
Gills: At first crowded but finally distant. Adnate, adnexed or broadly decurrent with age. Thin and obtuse at cap margin. White but with age becoming pale ochraceous from spores.

229 Lactarius turpis

230 Lactarius fulvissimus

231 Lactarius scrobiculatus

232 Lactarius lignyotus

Spores: Bright ochre in the mass, almost globose. 9–11 × 8–10 μm.
Stipe: Equal but may be enlarged at base, furrowed at apex, solid and dry. Velvety and concolorous with cap, except base which is whitish.
Flesh: Brittle at junction with the stipe, otherwise fairly elastic. White but bruising slowly pinkish, brownish with age. Taste may be mild or slightly acrid.
Milk: Copious at first but scanty with age. White and watery, mild or slightly acrid.
Habitat and season: Gregarious. Often found in damp areas under conifers. A beautiful species of N. America, far less common in Britain.
Edibility: No.

233 Lactarius torminosus (Fries) S. F. Gray SAID TO BE POISONOUS
Lactarius torminosus var. *sublateritius* Kühner & Romagnesi
Woolly milk-cap or Griping toadstool

Cap: 5–10 cm. Woolly and viscid at first. Convex then expanded and distinctly depressed. Margin involute and very shaggy with woolly fibrils. Pale flesh colour sometimes with an ochraceous tinge and often it is pale or even quite pink. Usually lightly zoned but not always so.
Gills: Slightly decurrent, thin and crowded. More or less concolorous with cap but usually paler. The parasitic Ascomycete—*Hypomyces torminosus*—is often present.
Spores: White and broadly elliptical. Ornamented with a network which can only be discerned under a high powered microscope. 7·5–9·5 × 6–7 μm.
Stipe: Cylindrical, more or less equal and clothed with minute depressed down. Internally pithy then hollow. Concolorous with or paler than cap.
Flesh: Firm in cap. White or cream-coloured. Taste acrid but with a pleasant smell.
Milk: White and unchanging. Abundant and acrid.
Habitat and season: Very common on heaths, or in grass under frondose trees. September to November.
Edibility: Said to be *poisonous* even *deadly*, but in Eastern Europe and Asia it is preserved in salt and eaten later seasoned with oil and vinegar. *We do not recommend.*

* **Lactarius controversus** (Fries) Fries

Similar to *L. torminosus.*

Cap: 5–14 cm. Dotted or zoned with blood-red spots.
Gills: Very crowded with intermediates.
Spores: Rough and globose. 6–8 μm.
Habitat and season: Found in woods and grass near trees especially poplar. From September to November.
Edibility: Yes.

* **Lactarius pubescens** (Krombholz) Fries

Again similar to *L. torminosus* but smaller in stature and more slender in habit.

Cap: Paler, not zoned and margin less woolly.
Spores: Rough and globose. 7–8 μm.
Habitat and season: Found in woods and grassland during autumn.
Edibility: No.

234 Lactarius blennius (Fries) Fries

Cap: Usually olive-grey, concentrically zoned with darker spots or broken lines, particularly towards the outer edge of cap. It is 5–10 cm across and convex, depressed in the centre and often (but not always) very slimy. The margin is incurved and somewhat downy.
Gills: White, bruising dirty grey. They are narrow, crowded and decurrent.
Spores: Pale buff in the mass. 7·5–8(9) × 6–7 μm.
Stipe: Robust and less in length than diameter of the cap. Slimy and similar in colour to cap.
Flesh: White.
Milk: Also white but turning grey on exposure to air.
Habitat and season: A common species found in deciduous woodland from August to November.
Edibility: Best avoided, very acrid.

* **Lactarius circellatus**

May be found under hornbeams. It also has a zoned cap but yellow gills.

* **Lactarius glyciosmus** (Fries) Fries
Coconut-scented milk-cap

Cap: 2–6 cm. Convex then plane, finally depressed with central papilla. Greyish-lilac and often rough or scaly.
Gills: Crowded and decurrent. Flesh colour to yellowish and later tinged greyish-lilac.
Spores: Creamy and almost spherical. 7–8 × 5·5–6·5 μm.
Stipe: Downy and solid. Paler than cap.
Flesh: Pallid with a strong odour of coconut.
Milk: White, rarely greenish. Unchanging. Mild at first then slightly acrid.
Habitat and season: Common in woods generally. August to November.
Edibility: Yes.

* **Lactarius vietus** (Fries) Fries

Somewhat similar to *L. glyciosmus* but its white milk turns grey on exposure to air or within thirty minutes of exposure. Has no coconut smell and very acrid taste.
Spores: Broadly elliptical. 8–9·5 × 6–7 μm.
Edibility: No.

Order-GASTEROMYCETALES

Having fruit-bodies where the hymenium is enclosed within a continuous, one- to three-layered peridium until the spores are ripe.

235 Endoptychum agaricoides Czernajev

Fruit-body: 1–7 cm wide and 2–10 cm high. Adpressed, fibrillose. Becoming dingy to pale tan and at times scaly.
Spores: Snuff-brown. 6·5–8 × 5·5–7 μm.
Flesh: Whitish, pale brown at maturity. Not or only slightly powdery.
Habitat and season: Scattered or gregarious on waste land, pastures and lawns etc. Summer and autumn in N. America. Not recorded in Britain.
Edibility: Good when young.

233 **Lactarius torminosus**

234 **Lactarius blennius**

235 **Endoptychum agaricoides**

236 **Mutinus caninus**

⋆ **Endoptychum arizonicum** (Shear & Griffiths) A. H. Smith & Singer

A closely-related species with spores 8·5–15 × 4·5–7 μm. Found in South-west of N. America. Not recorded in Britain.

236 Mutinus caninus (Persoon) Fries
Cyanophallus caninus (Huds.) Fries
Dog stinkhorn

Cap: Reddish-orange in colour and tapering upwards to a point. At first it is covered by a greeny-brown spore mass which has a slightly foetid smell.
Spores: Pale yellow and oblong. 4–5 × 1·5–3 μm.
Stipe: Porous, hollow and pitted. It varies in colour, sometimes white but often flushed with pink or even tinged orange.
Volva: White or greyish-white.
Habitat and season: Grows on humid ground in woodlands amongst dead leaves or around the dead stumps of trees. It develops from a small egg of about 2 cm long. See cross-section of egg in the illustration. A widespread species found from July to November.
Edibility: No.

Similar in size is the tropical species *Mutinus bambusinus* which has a red cap that extends half way down the stipe. This Phalloid has a strong foetid smell.

237 Phallus impudicus Persoon
Common stinkhorn or Wood witch

Cap: White, thimble-shaped and honeycombed, but covered at first by a blackish-olive, slimy spore mass. This mass has a most offensive and disgusting smell; it attracts many flies which feed thereon and so help in the distribution of the spores.
Spores: 3–5 × 2 μm.
Stipe: White, hollow and spongy standing 10–20 cm high and about 2½ cm thick.
Volva: Yellowish-white and completely envelopes the unripe mushroom giving the appearance of a hen's egg and about the same size and shape.
Flesh: Porous, fragile and white.
Habitat and season: Can be found in woods and gardens particularly under the fronds of bracken, when it is more readily located by its very unpleasant stench. Common from July to November.
Edibility: The 'egg' which is said to be edible can be found just beneath the surface soil attached to mycelial cords.

It was Fabre who said that photographic plates were affected by radiations emitted by this Stinkhorn, such radiations having passed through a cardboard box.

* **Dictyophora duplicata** (Bosc.) Fischer
Collared stinkhorn

A rare British species similar to *Pallus impudicus* but having a white net-like collar hanging from the underside of its cap, to halfway down the stipe. Also found in N. America.

* **Phallus hadriani** Persoon
Phallus iosmus Berkeley

Very similar to *Phallus impudicus* but differs in the colour of the 'egg' which in *P. hadriani* is violaceous.

Cap: Broader at the top with a much more distinct central rim. The net-work on the cap is usually more coarse and the stipe sturdier and usually larger in dimensions.
Spores: Spore mass is olive-brown. 3–5 × 1·5–2 μm.
Flesh: The odour is not as offensive as *P. impudicus*, sweeter.
Habitat and season: May be found in sand-dunes from July to November. Also recorded from gardens on the continent and in sandy areas in Czechoslovakia. Usually growing amongst *Ammophila* Host. (Marram Grass) and *Elymus* Linn. (Lyme-grass), perhaps in old rhizomes.
Edibility: No.

Genera-*Langermannia, Calvatia* and *Lycoperdon*

The fruit-bodies are globose or pear-shaped with a soft two-layered peridium.

238 Langermannia gigantea (Persoon) Rostkovius
Lycoperdon giganteum Batsch. ex Persoon
Calvatia giganteum (Batsch. ex Persoon) Lloyd
Giant puff-ball

Fruit-body: 5–50 cm in diameter but occasionally much greater. Globose and usually slightly depressed, smooth and chamois leathery, corrugated at base. Whitish becoming yellowish, then olive-brown and cracking. Has little or no sterile base; it is attached to the ground by a short root-like cord.
Flesh or Gleba: Firm and white at first, then yellowish, finally olive-brown and pulverulent.
Spores: Brown, finely warted and spherical. 4–5 μm. One fruit-body produces billions of spores released through cracks which start at the top.
Habitat and season: Locally common, usually gregarious in pastures, woods, roadside verges and gardens etc. An unmistakable species. (Illustration shows one of the authors holding a typical specimen.) They are often found year after year in the same spot.
Edibility: Very good whilst flesh remains firm and white.

239 Calvatia excipuliformis (Persoon) Perd.
Lycoperdon excipuliforme Persoon
Lycoperdon saccatum (Vahl.) Morgan
Elongate puff-ball

Fruit-body: 8–20 cm high with the head 2–10 cm across. Pestle-shaped. The cap peridium is whitish and covered with small needles or granules which fall off later. It then turns brownish and the upper part finally peels off. The base is sterile.
Flesh or Gleba: At first firm and white, then yellowish and later olive-brown, becoming pulverulent.
Spores: Olivaceous brown, warty and round. 4–5 μm. They are released through a crater-like hole in the top of the peridium.
Habitat and season: Occasional in woods, heaths and pastures. From July to November.
Edibility: Good so long as the flesh is firm and white.

240 Calvatia booniana A. H. Smith

Fruit-body: 20–60 cm broad and 7–30 cm high. Globular but somewhat flattened. Outer peridium is thick and white, sub-floccose and soon cracking into 4–6-sided patches about 2 cm deep. Finally eroding to expose a whitish then pale buff inner peridium. The fungus is attached to the substrate by a cord-like rhizomorph.
Flesh or Gleba: Whitish at first, dingy olive when mature and olive brown when dried.
Spores: Brown, globose to sub-globose or broadly elliptical. Smooth to very finely punctate. 4–6(6·5) × 3·5–5·5 μm.
Habitat and season: On bare ground under sage brush or in grassy places of arid regions. In the U.S.A. including Idaho, Oregon, Utah and Colorado, during late summer

237 **Phallus impudicus**

238 **Langermannia gigantea**

239 **Calvatia excipuliformis**

240 **Calvatia booniana**

and autumn. Not recorded in Britain.
Edibility: Good before the commencement of spore formation.

N.B. Our illustration is of a hand-painted plaque presented to Eric Soothill during a visit to Idaho where he was a guest of the North American Mycological Association and took part in the Wm. Judson Boone Foray held at McCall in 1976.

* **Calvatia cyathiformis** (Bosc.) Morgan
Bovista lilacina Berk & Mont.
Lycoperdon lilacinum (Berk.) Massee

Fruit-body: 7–16 cm in diameter and 9–20 cm high. Globose to turbinate, tapering abruptly into a large well developed thick stout rooting base, often wrinkled from base to half-way. Outer peridium is smooth or floccose, very thin and fragile, white to pale brown eventually cracking and falling away to expose the thin and delicate inner peridium. The latter erodes or breaks up along with the outer peridium, from the apex downwards, exposing the purple-brown gleba. Finally only the base is left, forming a cup containing fibrils and spores.
Flesh or Gleba: White then yellow and finally purple-brown.
Spores: Purplish-brown, globose to sub-globose, echinulate to echinate. (3·5)4·5–7·5 μm in diameter.
Habitat and season: Very common in grassy places in most parts of the U.S.A. and Canada during late summer and autumn. Not recorded in Britain.
Edibility: Good whilst the flesh remains white.

241 Lycoperdon perlatum Persoon
Lycoperdon gemmatum Batsch.
Common puff-ball

Fruit-body: 3–5 cm across and 4–8 cm high. Pear- or top-shaped with a small umbo. It is densely covered with short fragile spines each surrounded by a ring of smaller more numerous persisting warts. The spines soon erode leaving that part of the fungus with a net-like surface. White or dirty white, becoming yellowish-brown.

The stem-like base is usually free of spines and warts, but may be granular. It is sterile and coarsely cellular inside. White becoming yellowish-brown.

Spores: Olive, smooth and spherical. Released through an apical pore. 3·5–4·5 μm.

Flesh: At first white and firm with a pleasant taste and odour. Later it turns greenish-yellow and becomes soggy. Finally olive brown.

Habitat and season: Common and usually gregarious in open woodland or parkland. From July to December.

Edibility: Very good raw or cooked, but only while the flesh remains white.

242 Lycoperdon pyriforme Persoon
Pear-shaped puff-ball or Stump puff-ball

Fruit-body: 3–4 cm high and 2–4 cm broad. More or less pear-shaped, but occasionally rather umbonate. The outer peridium, which is yellowish, thin and flaccid, is covered with minute reddish-brown pointed warts. These erode with maturity leaving a smooth yellow or mealy surface.

Flesh: White, then greenish-yellow, finally olive to brownish as the spores develop. Has a central sterile column The spores are released through a small torn mouth at the apex. The base is furnished with white rooting strands of mycelium.

Spores: Olive, smooth and spherical. 4–4·5 μm.

Habitat and season: Common on decaying wood or on the ground but always attached to wood, usually densely tufted. August to November.

Edibility: Good whilst the flesh remains firm and white.

Genus-*Geastrum*-Earth-stars

The fruit-body is at first globe- or onion-shaped. The wall consists of three distinct layers; the two outer form the exoperidium, the innermost layer the endoperidium which is thin and papery, and sessile or stalked. Eventually the outer peridium splits from the apex downwards assuming a more or less star-like form. The inner globose peridium is left intact with an apical orifice.

All are terrestrial species and not edible.

243 Geastrum triplex Junghuhn
Collared earth-star

Fruit body: At first onion-shaped, 3–5 cm in diameter, and pale brown in colour. Later splitting radially from the top into a star like pattern 6 to 10 cm across of 5 or more thick fleshy rays, the upper sides of which are very light brown becoming darker with age. These ray-like segments often become cracked across and turn under the fruit-body giving an arched effect.

Endoperidium: Pale brown and sessile, surrounded at the base by a collar.

Spores: Dark brown, spherical and echinulate. 5–6 μm.

Habitat and season: Locally common in frondose woods, especially beech, from August to October. It is also found near the coast on the sandy slopes of conifer plantations in Lancashire, England.

Edibility: No.

* **Geastrum nanum** Persoon

Is much smaller than *G. triplex*, only 1·5–3 cm across, and has an outer peridium that splits into 5 to 8 pointed irregular rays. The inner umbonate sphere is stalked. Locally common on sea dunes and other sandy places from September to November.

* **Geastrum fornicatum** (Hudson) Fries

The largest species in Great Britain which stands from 5–12 cm high when fully opened and supported vertically on its 4 acute rays, the tips of which stay adhered to the substratum. The umbonate sphere is stalked and has a dark brownish appearance when covered with the ripe spores. A rare species which has an almost human profile and grows in grass at the edges of woodland.

244 Geastrum rufescens Persoon
Geaster fimbriatum Fries

Fruit-body: 2–3 cm in diameter, when expanded 3–6 cm across. Spherical at first. The rays are 6 to 8 in number, unequal and bluntly pointed, lying flat or recurved below the fruit-body.

Exoperidium: The outside surface is pale brown, the inner surface being yellowish watery-brown.

Endoperidium: Sessile, sub-ovate and about 1·5 cm in diameter. It is smooth and pale or umber-brown with a small apical pore.

Spores: Dark brown, spherical and warted. 4–6 μm.

Gleba: White at first, then ochre or olive and finally purplish-brown.

Edibility: No.

241 Lycoperdon perlatum

242 Lycoperdon pyriforme

243 Geastrum triplex

244 Geastrum rufescens

Genus-*Scleroderma*

The peridium is one layer, thick and tough. The spores are set free on decaying of fungus, usually commencing near apex.

245 Scleroderma verrucosum Persoon
Warted devil's snuff box.

Fruit-body: 4–8 cm across and flatish-ovoid in shape. Greyish to light brown. Its thin skin is finely warted at first, soon becoming smoother. Eventually it cracks irregularly to liberate the spores.
Spores: Dark brown with prickles. 10–13 μm.
Stipe: Absent but has a thick lacunose rooting base.
Habitat and season: Can be found growing on rich soil in frondose woods, also on waste ground and railway embankments. A not too common species from July to November.
Edibility: Not really edible although it is passed off, in cheap continental restaurants, as a truffle or used to adulterate them. Young specimens are sliced and sometimes darkened for this deceit. The taste bears no resemblance to that of a true truffle.

246 Scleroderma citrinum Persoon
Scleroderma aurantium Persoon
Scleroderma vulgare Fries
Common earth-ball

Fruit-body: 4–8 cm. Roughly oval in shape but often uneven. Hard and sessile, arising from cord-like mycelium.
Peridium: Thick, externally olive-yellow to brownish, scaly often in a net-like manner. In section whitish to rose-pink.
Gleba: Greyish, then purplish-black and veined whitish.
Spores: In mass are black with a purple tinge, eventually escaping through cracks in the peridium. Blackish-brown, reticulated and globose. 12–16 μm.
Habitat and season: Singly or gregarious under trees, especially frondose species. Very common from August to December.
Edibility: Worthless. Unlawfully passed off as truffles in certain countries.

Genera-*Cyathus, Crucibulum* and *Nidula*
Bird's nest fungi

The peridium is one- to three-layered.

247 Crucibulum laeve (de Candolle) Kambly
Crucibulum vulgare Tulasne
Cyathus crucibulum Persoon
Common bird's nest

Fruit-body: First seen as a somewhat globular body of greyish or a dirty cinnamon colour and if inspected closely will be found to have a double wall. Later it assumes the shape of an inverted bell or a basin, with a flat yellow membranous cover. On maturity this membrane bursts and disappears when we see its contents are numerous lens-shaped peridioles, dirty white in colour and each attached by a tiny cord (funiculus) to the inner wall of the bell-like receptacle.
Spores: Smooth. 7–10 × 3–5 μm.
Habitat and season: Although it is a common fungus growing in clusters on both frondose and coniferous sticks, wood chippings and the like, because of its small size (0·5–1 cm high and the same across) it is quite often difficult to locate. It occurs from September to March.
Edibility: No.

* **Cyathus striatus** Persoon
Striate bird's nest

Larger than *Crucibulum laeve* and has a greyish interior to its inverted bell with vertical ridges and containing 10–12 greyish eggs. It also grows in a similar habitat but occurs from March to November. Common.

* **Cyathus olla** Persoon

Very similar in appearance to *C. striatus* but the bell-shaped sporophore tapers more towards its base and the inside is smooth. Habitat also the same and occurring March through to November. A fairly common species.

248 Nidula candida (Peck) White
Bird's nest fungus

Fruit-body: 1–1·5 cm high. Single-walled, rough and shaggy with a wide flaring mouth. Grey or light brown in colour. Becoming bleached white with age. The peridioles are large, without stalks and some 1·5–3 mm across. They are grey or light brown and with a thin tunica.
Spores: Hyaline and ellipsoid. 8–10 × 4–6 μm.
Habitat and season: Gregarious on fallen debris in parts of N. America. March to November. Not recorded in Britain.
Edibility: No.

245 Scleroderma verrucosum

246 Scleroderma citrinum

247 Crucibulum laeve

248 Nidula candida

Order-TREMELLALES

249 Hirneola auricula-judae (St. Amans) Berkeley
Auricularia auricula (Hooker) Underwood
Jew's ear fungus

Fruit-body: 3–10 cm across. Ear-shaped or very irregular cup-shaped. Tough, translucent and gelatinous in wet weather. Hard and shrivelled if dry weather persists. Outer surface is velvety with greyish hairs.
Hymenium: Smooth, veined and reddish to purplish-brown.
Spores: White, oblong, curved and narrow at the base. 16–20 × 5·5–8·5 μm.
Habitat and season: Tenaciously attached to the branches of frondose trees, mainly elder but also elm, beech, walnut, willow and pines in Western North America. May be found at any time of the year. A common species.
Edibility: Said to be good and can be eaten raw in salads. In China a closely related species is regarded as a delicacy, this fungus is cultivated and grown on lengths cut from oak saplings which are left to lie on the ground until infected with the mycelium of the fungus.

250 Tremella lutescens (Persoon) Fries
Witch's butter

Fruit-body: 1–5 cm broad. Very soft and tremulous. Gelatinous and brain-like with crowded lobes. Naked, becoming hoary with the spores. Pale whitish-yellow then yellow.
Spores: Hyaline and sub-globose. 12–15 μm in diameter.
Habitat and season: Common on fallen branches and stumps from August to November.
Edibility: Not known.

* **Tremella frondosa** Fries

This is a large species up to 10–20 cm high and broad. With undulate and contorted lobes of pale pinkish-yellow. Found rarely on the trunks of frondose trees, especially oak. Spores are sub-globose. 7–9 μm. July to December.

* **Tremella mesenterica** Hooker
Yellow brain-fungus

This is a bright orange-yellow species up to 10 cm broad. Gelatinous but firm with short lobes. Smooth, and pruinose. Spores are white, smooth and broadly elliptical. 7–10 × 5–8 μm. A common fungus on dead branches and sticks. From January to December.

251 Exidia recisa (S. F. Gray) Fries
Witch's butter

Fruit-body: 1–3 cm broad. Hemispherical or disc-shaped and rough with dots. Brownish-yellow, gelatinous and translucent.
Spores: White, cylindric and curved. 15–18 × 4–5 μm.
Stipe: Short, eccentric and concolorous with fruit-body.
Habitat and season: Fairly common, usually growing in clusters on the branches of willow. September to December.
Edibility: No.

* **Exidia glandulosa** (St. Amans) Fries

Blackish and gelatinous with a contorted brain-like appearance and rough with dots.
Fruit-body: 2–6 cm across, often in crowded groups on stumps and fallen branches of frondose trees.
Spores: White, cylindrical and curved, 12–15 × 4–5 μm.
Stipe: Short.
Habitat and season: Can be found all the year round.
Edibility: No.

252 Calocera viscosa Fries
The beautiful horn or Sticky coral fungus

Fruit-body: This species grows from 3–8 cm high. Of an attractive golden-yellow and reminiscent in shape of stags' horns. In dry weather they are appreciably darker in colour. The horns are soft and gelatinous outside but have a firmer more substantial interior.
Spores: Ovoid and yellow. 8–11·5 × 3·5–4·5 μm.
Habitat and season: Only found on conifer stumps or roots, to which they are tenaciously attached. If climatic conditions are suitable can be found throughout the year.
Edibility: Worthless.

* **Calocera cornea** Fries

Fruit-body: About 1 cm in height. Club-shaped and unbranched. Yellow in moist atmosphere, orange when dry.
Spores: White and arcuate. 7–9 × 3–4 μm.
Habitat and season: Common and gregarious on damp wood, especially beech. May be found throughout the year.
Edibility: No.

* **Calocera glossoides** Fries

Fruit-body: Between 1 and 1·5 cm high. Simple and unbranched. Pale yellow and somewhat tremellose. Round and slender at base, somewhat flattened above.
Spores: (9·5)12–14(17) × 3·2–4(4·75) μm.
Habitat and season: Common on damp wood, fallen branches etc. Found most months of the year.
Edibility: No.

249 Hirneola auricula-judae

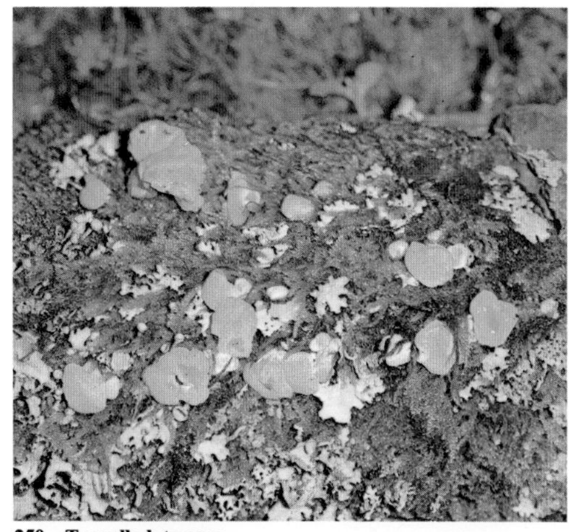

250 Tremella lutescens

251 Exidia recisa

252 Calocera viscosa

Bibliography

Bertaux, A. (1966). *Les Cortinaires*, Paris.
Blum, J. (1962). *Les Bolets*, Paris.
Blum, J. (1962). *Les Russules*, Paris.
Brodie, Harold J. *The Bird's Nest Fungi*, University of Toronto Press.

Christensen, Clyde M. *Common Edible Mushrooms*, University of Minnesota Press.
Cooke, M. C. (1880–90). *Illustrations of British Fungi*, London.

Dennis, R. W. G. (1954). *The Genus Inocybe*, 'The Naturalist', London.
Dennis, R. W. G. (1954). *The Genus Mycena*, 'The Naturalist', London.
Dennis, R. W. G., Orton, P. D. and Hora, F. B. (1960). *New Check List of British Agarics and Boleti.*

Essette, H. (1964). *Les Psalliotes, Atlas Mycologiques I*, Paris.

Harrison, Kenneth A. *The Stipulate Hydnums of Nova Scotia*, Canada Department of Agriculture.
Heim, R. (1957). *Les Champignons d'Europe*, Paris.
Hesler, L. R. (1967). *Entoloma in S. E. North America*, Nova Hedwigia, Beiheft.
Hesler, L. R. & Smith, A. H. (1963). *North American Species of Hygrophorus*, Knoxville, Tennessee.
Hesler, L. R. & Smith, A. H. (1965). *North American Species of Crepidotus*, New York.
Hora, F. B. (1957). *The Genus Panaeolus in Britain*, 'The Naturalist', London.
Hvass, E. & H. (1961). *Mushrooms and Toadstools in Colour*, London.

Kühner, R. & Romagnesi, H. (1953). *Flore Analytique des Champignons Supérieurs de France*, Paris.

Lange, J. E. (1935–40). *Flora Agaricina Danica*, Copenhagen.
Lange, M. & Hora, F. B. (1963). *Collins Guide to Mushrooms and Toadstools*, London.
Leclair, A. & Essette, H. (1969). *Les Bolets, Atlas Mycologiques II*, Paris.
Locquin, M. (1956). *Petit flore des Champignons de France*, Vol. 1, Paris.

Massee, G. (1892–3). *British Fungus Flora. Vol. I–IV*, London.

Miller, Orson K. Jr. (1975), *Mushrooms of North America*, E. P. Dutton & Co., Inc., New York.
Miller, Orson K. Jr. & Farr, David F. (1975). *An Index of the Common Fungi of North America*, J. Cramer.

Nicholson, B. E. & Brightman, F. H. (1966). *Oxford Book of Flowerless Plants*, Oxford.

Orton, P. D. (1955). *The Genus Cortinarius I, Myxacium & Phlegmacium*, 'The Naturalist', London
Orton, P. D. (1958). *The Genus Cortinarius II, Inoloma & Dermocybe*, 'The Naturalist', London.

Pearson, A. A. (1946). *British Boleti*, 'The Naturalist', London.
Pearson, A. A. (1950). *The Genus Lactarius*, 'The Naturalist', London.
Pearson, A. A. (1948). *The Genus Russula*, 'The Naturalist', London.
Pearson, A. A. (1955). *The Genus Mycena*, 'The Naturalist', London.
Pearson, A. A. (1954). *The Genus Inocybe*, 'The Naturalist', London.
Peterson, Ronald H. (1971). *The Genera Gomphus and Gloecantharellus in North America*, Verlag von J. Cramer.
Pilat, A. & Usak, O. (1951). *Mushrooms*, London.
Pilat, A. & Usak, O. (1961). *Mushrooms and other fungi*, London.
Pomerleau, R. & Jackson, H. A. C. (1951). *Mushrooms of Eastern Canada and The United States*, Les Editions Chantecleer, Montreal.

Ramsbottom, J. (1923). *A Handbook of The Larger British Fungi*, British Museum, Natural History, London.
Ramsbottom, J. (1943). *Edible Fungi*, London.
Ramsbottom, J. (1951). *Handbook of Larger Fungi*, London.
Ramsbottom, J. (1953). *Mushrooms and Toadstools*, 'New Naturalist', London.
Ramsbottom, J. (1945). *Poisonous Fungi*, King Penguin, London.
Rea, C. (1922). *British Basidiomycetae*, Cambridge.
Rinaldi, A. & Tyndala, V. (1972–74). *Mushrooms and other fungi*, Hamlyn, London.
Romagnesi, H. (1956–65). *Nouvelle Atlas des Champignons*, Bordas.
Romagnesi, H. (1963). *Petit Atlas des Champignons*, Bordas.

Romagnesi, H. (1967). *Les Russules d'Europe et d'Afrique du Nord*, Paris.

Singer, R. (1965). *Die Röhrlinge I, Die Boletaceae (ohne Boletoideae)*, Die Pilze Mitteleuropas.

Singer, R. (1967). *Die Röhrlinge II, Boletoideae und Strobilomycetaceae*, Die Pilze Mitteleuropas.

Smith, A. H. & Hesler, L. R. (1968). *North American species of Pholiota.*

Smith, A. H. (1947). *North American species of Mycena*, Ann Arbor, Michigan.

Smith, A. H. *A Field Guide to Western Mushrooms*, The University of Michigan Press.

Smith, A. H. & Singer, R. (1964). *A monograph on the genus Galerina*, Earle, New York.

Smith, A. H. & Thiers, H. D. (1964). *A contribution toward a monograph of North American species of Suillus*, Ann Arbor, Michigan.

Step, E. S. *The Harvest of The Woods*, Jarrold, London.

Von Frieden, Lucius (1969). *Mushrooms of The World*, The Bobbs Merrill Co. Indianapolis–New York.

Wakefield, E. (1954). *Observer's Book of Common Fungi*, London.

Wakefield, E. & Dennis, R. W. G. (1950). *Common British Fungi*, London.

Watling, Roy (1969). *British Fungus Flora (Agarics and Boleti)*, H.M. Stationery Office, Edinburgh.

Watling, Roy (1970). *Identification of the Larger Fungi*, Hulton, Amersham.

Wells, Virginia L. & Kempton, Phyllis E. *A Preliminary Study of Clavariadelphus in North America.*

Index

Scientific Names

Index

Common Names